To Eric —
Build Businesses!

Praise for *Inside the Minds*

"Unlike any other publisher – actual authors that are on the front-lines of what is happening in industry." – Paul A. Sellers, Executive Director, National Sales, Fleet and Remarketing, Hyundai Motor America

"What C-Level executives read to keep their edge and make pivotal business decisions. Timeless classics for indispensable knowledge." – Richard Costello, Manager-Corporate Marketing Communication, General Electric

"Want to know what the real leaders are thinking about now? It's in here." – Carl Ledbetter, SVP & CTO, Novell, Inc.

"Aspatore has tapped into a gold mine of knowledge and expertise ignored by other publishing houses." – Jack Barsky, Managing Director, Information Technology & Chief Information Officer, ConEdison *Solutions*

"Priceless wisdom from experts at applying technology in support of business objectives." – Frank Campagnoni, CTO, GE Global Exchange Services

"Aspatore publishes the answers to every business person's questions." – Al Cotton, Director, Nypro Corporate Image, Nypro Inc

"Everything good books should be - honest, informative, inspiring, and incredibly well-written." – Patti D. Hill, President, BlabberMouth PR

"Answers questions that others don't even begin to ask." – Bart Stuck, Managing Director, Signal Lake LLC

"Unique insights into the way the experts think and the lessons they've learned from experience." – MT Rainey, Co-CEO, Young & Rubicam/Rainey Kelly Campbell Roalfe

"Must have information for business executives." – Alex Wilmerding, Principal, Boston Capital Ventures

"Reading about real-world strategies from real working people beats the typical business book hands down." – Andrew Ceccon, Chief Marketing Officer, OnlineBenefits, Inc.

"Books of this publisher are syntheses of actual experiences of real-life, hands-on, front-line leaders--no academic or theoretical nonsense here. Comprehensive, tightly organized, yet nonetheless motivational!" – Lac V. Tran, Sr. Vice President, CIO and Associate Dean Rush University Medical Center

"Aspatore is unlike other publishers…books feature cutting-edge information provided by top executives working on the front-line of an industry." – Debra Reisenthel, President and CEO, Novasys Medical, Inc

www.Aspatore.com

Aspatore Books is the largest and most exclusive publisher of C-level executives (CEO, CFO, CTO, CMO, partner) from the world's most respected companies and law firms. Aspatore annually publishes a select group of C-level executives from the Global 1,000, top 250 law firms (partners and chairs), and other leading companies of all sizes. C-Level Business Intelligence™, as conceptualized and developed by Aspatore Books, provides professionals of all levels with proven business intelligence from industry insiders—direct and unfiltered insight from those who know it best—as opposed to third-party accounts offered by unknown authors and analysts. Aspatore Books is committed to publishing an innovative line of business and legal books, those which lay forth principles and offer insights that, when employed, can have a direct financial impact on the reader's business objectives, whatever they may be. In essence, Aspatore publishes critical tools—need-to-read as opposed to nice-to-read books—for all business professionals.

Inside the Minds

The critically acclaimed *Inside the Minds* series provides readers of all levels with proven business intelligence from C-level executives (CEO, CFO, CTO, CMO, partner) from the world's most respected companies. Each chapter is comparable to a white paper or essay and is a future-oriented look at where an industry/profession/topic is heading and the most important issues for future success. Each author has been carefully chosen through an exhaustive selection process by the *Inside the Minds* editorial board to write a chapter for this book. *Inside the Minds* was conceived in order to give readers actual insights into the leading minds of business executives worldwide. Because so few books or other publications are actually written by executives in industry, *Inside the Minds* presents an unprecedented look at various industries and professions never before available.

INSIDE THE MINDS

Biotechnology Venture Capital Investments

Leading VCs on Deal Terms, Valuations & Keys to Success for Investors and Management Teams

C-LEVEL BUSINESS INTELLIGENCE · C-LEVEL BUSINESS INTELLIGENCE

ASPATORE BOOKS

Inside the Minds Project Manager, Merissa Kraham; edited by Michaela Falls; proofed by Melanie Zimmerman

ISBN 978-1-59622-644-9
Library of Congress Control Number: 2007921071

For corrections, updates, comments or any other inquiries please e-mail editorial@aspatore.com.

First Printing, 2007
10 9 8 7 6 5 4 3 2 1

Biotechnology Venture Capital Investments

Leading VCs on Deal Terms, Valuations & Keys to Success for Investors and Management Teams

CONTENTS

Building Businesses With Venture Capital

Wm. Blake Winchell
Managing General Partner
Fremont Ventures

The "Duh" Factor

This chapter is about building businesses in biotechnology using venture capital. Many of you who read this will say, "Duh, tell us something we don't know." Others might say, "This is all very common sense stuff, and it's nice to be reminded once in a while," and I'm sure there's a group that will say "This guy doesn't have the faintest idea what he's talking about." I dedicate this chapter to that third group.

Venture capital funds are invested and managed as portfolios, with some funds being diversified by investing in both IT and life sciences. There is often further diversification (even within a life science-only fund) by investing in medical devices, diagnostics, platform companies, drug development companies, and sometimes health care information technology companies. There is also differentiation between investments in how long they will be held by the fund. Some (mostly later-stage investments and service-oriented businesses) may be held for as little as two to five years, and others, often earlier stage investments in therapeutic development companies, may take up to ten years, which is the life of a typical venture capital fund.

These companies look very different from one another in terms of their "business models" (see below), the kind of entrepreneurs who will be successful in running them, how much venture capital they will require, and what valuation they might ultimately achieve for their employees and investors. Service and device companies may generate revenue early and require less venture capital over a shorter period of time before they are "monetized," while a successful drug development company may take almost a decade and well over $100 million of investment before it generates value for its employees and investors.

Not surprisingly, the successful CEOs of these companies also have different characteristics. Companies that need to raise hundreds of millions of dollars prior to having a product tend to be run by individuals who are very, very good at raising money. Successful businesses that have products early and need to generate revenue are most often run by teams that are excellent at product development, launch, marketing, and sales.

Ultimately, however, all of these enterprises have one thing in common—they must at some point be a "business." Whether it takes two years or ten years, the enterprise must generate a product or service of higher value to customers than the costs required to build it, or it will fail.

A balanced life sciences portfolio may contain a couple of "swing for the fences" drug development companies, as well as several opportunities which require less capital and are expected to generate positive but more modest returns. The rest of this chapter focuses on the latter type of investments, and in many ways represents an extreme position in order to make the point "It's all about the business." In the past, venture capitalists have often been able to generate excellent returns by "playing offense"—aggressively funding companies and then pushing them out into the public markets. The new rules of the game may better reward a more defensive approach to biotech VC deals.

It's All about the Business

Over the past several years, the venture capital industry has changed dramatically and permanently. The days when venture investors were able to sell a company that consisted of a hot technology, a prototype product, or an unproven drug for a billion dollars are over—let alone those days gone by when we could take such a fledgling company public at a huge valuation. It's very easy to make a venture capital investment—simply write a check. It is significantly more difficult to monetize that investment through a profitable sale or IPO.

For many life sciences deals that means you must build a real *business*, not just a company, in order to create true value in the future. A business sells a product or service that customers demand. This generates revenue. As venture investors we are always asking ourselves "How big a business can we build?" Size of market, product performance, and management's ability to build a business are what we invest in. By comparison, a company is a collection of people. It is easy to build a company—all you have to do is hire people. But building a business requires creating an offering that has value and can be built into an economically valuable enterprise—it is not a single product, feature, or platform, but a true business. This means an organization must have scale, sustainable profits, and competitive barriers (often intellectual property).

In such a "business model," a company must behave like a business, which involves treating cash as a scarce resource and scaling infrastructure behind revenue, not ahead of it. It also means structuring business strategy to generate revenue early in the company's life cycle and focusing steadily on the one or two things that are going to make this business successful.

Referring to "biotech investing" is similar to saying "real estate investing." Just like investing in one (or many) multi-family residences is worlds away from developing a huge shopping center, life sciences investments come in all shapes, sizes, and flavors, some of which are more conducive to generating early revenues than others. Generating early revenues is a more straightforward objective for a medical device, diagnostic or technology platform company than for a therapeutic development company. However, even companies with very long development cycles (such as drug discovery companies) requiring tens or even hundreds of millions of dollars in funding can generate cash from partnering, grants, and co-development that will reduce the amount of venture capital required.

As difficult as successful venture investing has been in the past, it will be even more difficult in the future. It will involve finding businesses that offer true visibility into the timing and size of revenue streams, and it will also mean focusing on investments where the company will not need $100 million in capital just to reach the point where it becomes clear whether or not the drug or device will work. Responsible venture capital investing will also mean making fewer investments and working diligently at the board level to make sure that management is making the right decisions. Ultimately, success will mean not being afraid to change management or strategy if you find that your original team or investment thesis is flawed or not working.

Greed and Fear in Investing

There is an old adage about greed and fear: opportunity exists when entities with a lot of money are either very greedy or very afraid. These words remain true in the biotech industry today, whether you're dealing with doctors looking for ways to bring more procedures into their offices or with hospitals seeking to wring every last dollar from the reimbursement codes. It is just as true when dealing with pharmaceutical companies that are afraid of their empty pipelines or generic competition; likewise, it is relevant when dealing with consumers who lust for a better sex life or are afraid of getting old, fat, or gray (or, heaven forbid, all three!).

Successful ventures almost always require people to change their behavior, something that does not come easily. It takes a powerful motivator to compel an individual to change the way he or she does things, such as affecting the way a doctor prescribes a drug or performs a procedure. The same holds true when seeking to impact the behavior of a pharmaceutical company that has pursued the same process of drug discovery and development for twenty years. Identifying a need is not enough. You have to change behavior, and, as investors and entrepreneurs, we almost always underestimate how difficult that is. It is often said about start-ups that a ten percent productivity improvement for customers will not convince them to change, but a tenfold improvement always will. Businesses that actually deliver tenfold improvements are few and far between.

Another approach when seeking investment opportunities is to look for seams of change in large sectors and identify the investments that will benefit from those changes. The often-discussed aging of the Baby Boom generation has created huge opportunities, such as those related to vision correction or a new drug targeted at slowing the loss of hearing.

Venture Capital as a Scarce Resource

At the most basic level, everyone in this industry is approaching biotech venture capital in the same way, which is to scour every nook and cranny to invest in the best management teams developing the most valuable technologies. We expect that these teams will be dedicating themselves to businesses that have products and/or services that satisfy a compelling need and can be financed with significantly less capital than the price for which they will ultimately be sold.

Investors who are sufficiently confident in their own judgment regarding the ultimate success of an investment have pursued, sometimes successfully, a strategy of overpowering the competition with resources (read cash). This strategy, however, is not for those with faint hearts or thin wallets.

A more contrarian way to approach biotech VC investing is with a "scarcity mentality" rather than an "abundance mentality." With a scarcity mentality, investment capital is scarce, not bountiful. In contrast to this, I have actually heard an investor at a board meeting say, "Let's hyper-finance this company and own this space before anyone else has a chance." Our response to this approach is, to quote Monty Python, "Run Away! Run Away!"

Our belief is that the cheapest, least dilutive way to finance a company is with gross margin; this means finding customers or partners who are willing to write checks—preferably large ones—to access the company's products or services early in the company's life cycle. It also means waiting to build infrastructure until those customers are secured, even though that strategy is sure to stretch both human and financial resources to their breaking point. It does not mean being penny-wise and pound-foolish, but (to slightly mix the metaphor) we shouldn't be talking with Benjamin when Hamilton will do.

Management teams are often overly confident that their venture is going to be a winner; they are so sure that growth is imminent and they know exactly what the market needs or how a compound will perform, they want to put the infrastructure in place as soon as possible. This "build-it-and-they-will-come" mentality invariably leads to two outcomes: First, since revenue always takes longer to materialize and ramps more slowly than expected, the infrastructure burns precious cash as it awaits customers to serve; second, what customers want and are willing to pay for is invariably different (often wildly so) than what the company has built.

Our approach is also a "pay-as-you-go" investment strategy. Instead of giving a company three years worth of capital up front, we would rather provide it with twelve to eighteen months of cash under a conservative business plan with cautious burn rate assumptions. If the investment is not getting the kind of market acceptance we would expect, we can then adjust the go-to-market strategy and/or the management team early in the process. Some investors might say that this financing strategy consumes precious management time in fundraising that could be better spent. However, our belief is that having to go back to old investors or seek new ones every year or two is a valuable reality check on the business' progress.

At the same time, we believe in supporting our businesses that are meeting their milestones, even as we require them to bring new investors into the next round. Current investors will offer to put up a third to a half of the next round if the team can bring in one to two new investors for the balance. It is also very important that, if the team is going to spend the time to raise another round of funding, it is large enough to bring the business to sustainability. That doesn't mean that the business will never require additional capital, for once it has demonstrated its viability, both the investors and management may decide to invest additional "growth" capital. Ideally, though, once the team has validated the business model with the first round of funding, the second round should be enough to bring it to breakeven or, at a minimum, to a major value milestone using very conservative assumptions.

The days where companies could lose money for six or seven years straight and continue going back to the well are numbered. The mantra of the 90s, "get big fast," has turned into the mantra of the new millennium, "grow fast cheap." Outsourcing non-core tasks (like HR or IT management), turning fixed costs into variable costs (like making compensation for business development highly bonus dependent), partnering early-on with firms with complementary expertise (like partnering the medicinal chemistry for a development program rather than hiring the resources), and most importantly, leveraging marketing and sales through partners (rather than trying to do everything direct) can lower a company's burn rate and preserve cash. The last thing this world needs is another 'vertically integrated specialty pharmaceutical company'.

In much the same way as calorie restriction has been shown to improve health and retard aging in a variety of animal species, a scarcity mentality when applied to venture capital investing can do the same in a variety of business species, including biotechnology. An organism that is given a high calorie diet develops a high rate of metabolism and cellular fat. Similarly, a company that is provided with a great deal of cash usually will develop an infrastructure sufficient to metabolize that cash. Current views propose that calorie restriction places an organism in a defensive state so that it can survive adversity, and this results in improved health and longer life. A business that is bootstrapped with scarce cash will, by necessity, direct its limited resources to the most strategically critical areas, staying lean, mean, and focused.

The Art (and Luck) of Venture Investing

In truth, the most important factor in a venture capital investment is luck. Ask any successful venture investor, and, if he or she is honest with you, you will hear that most of their money has been made as a result of getting lucky—that is, by being in the right place at the right time with the right investment. You also need to be investing in a portfolio of businesses, because in venture capital many of your investments will not work out. If you are lucky, the ones that do work out are so wildly successful that they return significantly more money than is lost on the ones that fail.

Many VCs made a lot of money in 1999 and 2000 on IPOs of companies that are no longer with us. From Dr. Koop to CommerceOne, for a few golden months, the stars lined up for those investments to make their backers billions of dollars. Of course, hard work usually improves one's luck substantially, but luck and timing play a larger role in successful venture investing than most people realize.

Even though we are investing in a field associated with science, anyone who has been in this business for more than a few days will tell you that there is more art than science to biotechnology. Those who are not involved in the field may assume that biotech venture capital investing is centered upon predictable, scientific rules, yet a significant portion of product development is based on trial and error where luck plays a major role. Much of business development and sales are based on being in the right place in front of the right person at the right time. All of these elements are outside of the company's control, and tapping into them perfectly is more of an art than a science.

The most challenging aspects of biotech venture capital involve knowing when to be patient, sticking with the strategy, managing costs aggressively, and either cutting programs appropriately or making a commitment to invest more financial and human resources ahead of revenue traction. Finally, it is critical to know when to say "that's enough," at which point you either sell the company or shut it down rather than investing good money after bad. If venture returns in the future are going to be composed of doubles and triples rather than "ten baggers" (e.g., generating returns of two or three times invested capital rather than ten times), one venture fund cannot afford to invest $30 million in a single company and eventually write the whole thing off.

The venture investors who have been most successful at practicing this art have a difficult time explaining what it is that makes them successful. Grant Heidrich (Millennium Pharmaceuticals, Tularik) calls it "the nose." Brook Byers (Idec Pharmaceuticals, Athena Neurosciences) says that the longer he spends in this profession, the more complex he realizes it is, and the more difficult it becomes to explain.

What successful biotech investors have in common (besides luck) is judgment—judgment about technology, judgment about people, about markets, and about investment timing. That judgment has been honed over many years and hundreds of decisions—many good and many mistakes.

How Much is Enough?

When investing in a new company, we look for companies that can reach $25 to $50 million in revenue during our holding period with the promise that getting to $100 million or more in revenue is realistic for an acquirer or public offering.

In evaluating our potential return on investment, we expect to make three to seven times (or more if we get lucky – see above) our total money invested in any single investment, recognizing that our return calculations, like those of all of the companies in which we invest, are virtually always optimistic. For every Esperion, a very early stage drug development company that was purchased by Pfizer for $1.3 billion, there are dozens of flameouts. An important element that we consider when entering into any investment is our exit: Do we expect the business will be an IPO candidate or will it likely be acquired? In the latter situation, we try to determine the profile of the most probable acquirers and attempt to position the company with them early on to spark their interest in the opportunity, while building a business that can also be attractive to other strategic partners and investors.

One of the greatest challenges in venture investing is knowing when to pull the plug and close down or sell for technology value a company that is non-performing versus when to double down on your investment. In any portfolio of ten companies, we know that several, if not most, will lose money. The problem is that we don't know which of the ten they are, because at the time of the initial investment decision, we obviously think this is a team and business model that will work. What that means is that the more we invest in companies that eventually fail, the more we have to make on the ones that are successful in order to generate the returns that we expect of ourselves and that are demanded by our limited partners.

For example, if a fund invests in ten companies over five years, and if we invest $1 million in each of them at an early stage, and three "flameout," and we invest $5 million each in the other seven, we will have invested a total of $45 million. If it costs a total of $20 million for a venture start-up to develop into a sustainable business, then we and other investors may have invested a total of $140 million invested in those seven companies. If only two of the remaining seven can exit by going public or being acquired, we need our ownership in those two companies to be worth $200 million in order for us to generate the returns our limited partner investors demand (20%+ IRRs). That means those two companies, in aggregate, must be worth about one billion dollars when we exit. While this may seem like an unrealistic example, you get the point. Very few venture investors consistently invest in ten start-ups that become two businesses with an average market value of $500 million each. The point here is that we need to preserve our capital for those start-ups that after some period of maturation look like they are worthy of additional investment. And even then, only a few will actually become successful.

Mitigating Execution Risk through Superior Management

Like most people in this business, I have learned about risk the hard way: by making bad decisions and having to live with the consequences. The biggest risk that any venture capital-backed company faces is execution risk:

can the team deliver the plan within the stated time frame for the budgeted funding? In real estate, the ticket to success is "location, location, location"; in venture investing, the key is "execution, execution, execution." Clearly, there are a myriad of other risks, such as technology risk, competitive risk, and market risk, however execution risk trumps them all. The "first mover" advantage is nothing when compared to the "excellent execution" advantage.

How, then, do you mitigate this huge risk to success? In my book, there is no substitute for what we refer to as "experienced judgment" in a company's CEO. True experience means deep domain expertise, leadership and management skills, scientific vision, and documented success in convincing others to share that vision. Preferably this success translates to the ability to inspire others to write a check, whether as an investor or as a customer. When identifying judgment as an important trait in a CEO, one must assess the CEO's instinct to do the right thing in challenging scenarios. This instinct has almost always been honed by making errors in managing other companies, on other people's dimes, and then learning from those mistakes. We want to take advantage of that experienced judgment that others have paid for in the past by acquiring as CEOs those executives who have made—and learned from—prior mistakes.

We do not believe in on-the-job training for CEOs; the biggest risk a company can take is betting on someone who has never performed in this role before. The right CEO will be able to attract a strong team, and a strong, experienced team is more likely to avoid making company-killing mistakes. For every Steve Jobs there are a thousand failed first-time CEOs. Even Steve stubbed his toe with Next Computer before he succeeded as CEO of Apple.

When building a management team, the key quality we look for in a CEO is leadership. A great CEO will attract great scientists and will understand the importance of IP and exceptional science. Strong CEOs can persuade both prospective employees and customers to buy into "the vision" and then steer the company through the inevitable myriad problems to a successful commercial launch, growth, and, ultimately, liquidity event.

We view our role as investing in the CEO and then working with him or her to build a management team, consult on product development, strategy, winning customers, raising and investing additional capital, and helping the CEO succeed. In our opinion, this is the difference between venture capital firms and investors. Do you hire managers and work to help them succeed or do you hire managers and then fire them when they do not? The first is a venture firm, the second are just investors. That is not to say that we expect every CEO we hire to be the one that will take the company all of the way to a profitable exit. As a business go through different phases from start-up to maturity, the profile of the CEO who is best suited to run it at any given phase will be different.

We were proud to be recognized by *EE Times* as one of the "Ten Best VCs" in the U.S. for adding value to portfolio companies in a survey they did in October 2005. Many of the other in the "Top Ten" (Kleiner Perkins, Sequoia, NEA, US Venture Partners, Greylock, and Sevin Rosen) have been around decades longer than we have and have ten times the amount of capital under management. Despite being the youngest and smallest in the group, the article noted that "Fremont was cited for the quality of its support to its companies"; it's the highest praise we could have asked for. The point is not the survey itself, what's important is that there is a compact between good VCs and the companies they invest in. That compact requires not only money, but true support from the venture firms in return for performance by the company.

The Best of Intentions Lead to the Biggest Mistakes

The most frequent mistakes made in this industry are mistakes of the heart. Truly believing in your science or product is an admirable trait, but it is one that can lead to blindness, preventing you from seeing the problems plaguing the actual business. Examples of this type of error include failing to replace a member of the management team out of loyalty or hiring a less-than-stellar employee because that role desperately needed to be filled. Another related mistake is delaying cost reductions because you are positive that the big deal that you have been discussing is imminent or because you are confident that the venture firm you are in negotiations with is just days away from writing you a check. Sometimes a mistake of the heart will surface in the form of submitting a budget that demonstrates huge revenue and expense growth because of your confidence that the need is there; as a result, the management team ends up ramping expenses according to plan even though revenue is woefully behind this same plan. With very few exceptions, management teams of venture-backed companies are earnest, honest, and extremely hard working. The mistakes that they make, and that we as board members often sanction, are usually born of the best of intentions.

With respect to venture capitalists themselves, common mistakes, at least in our experience, are those that result from ignoring one's own judgment about investments. I have learned that if my gut tells me not to invest in a company, it is almost always right. If your antenna says that there is something off about a CEO candidate, do not make the offer, no matter how much you need to hire someone. A budget that feels unattainable usually is. Further, if a company's burn rate is too high, it must be reduced.

One of our portfolio companies was burning over $2 million per month in cash, and we were a lone voice on the board imploring the CEO to reduce headcount. He said 'all of the fat has been cut; there's nothing left but muscle and bone'. That company recently filed Chapter 11.

We have never replaced a CEO of any of our companies too early; in fact, one of my mentors is fond of saying, "If you think you have a problem with a CEO, you have already had it for a while."

Reputation and Returns

The two most important components for a good track record over the long term in this industry are returns and reputation. These two attributes will always converge. Returns are the profits that a venture firm generates from a portfolio of investments over five to ten years. Reputation is how a venture firm is regarded by other investors and entrepreneurs. Are they fair or are all of their terms one-sided? Do they add value at the board level or practice "seagull investor management" swooping in and leaving a trail of guano behind? Are they smart and strategic or is their input superficial and uninformed? Do they do their homework or do they read the board package for the first time during the meeting itself?

A firm with great returns and a poor reputation will find that, over time, one of those two things will change; most likely its returns will go down as it does not carry its weight in investments and is not invited into the most attractive deals. Often, the only way to change a poor reputation is to change the people in the firm. The same holds true for a firm with a great reputation and poor returns. Over time, returns are likely to improve as the value added manifests itself in their portfolio companies, and entrepreneurs and co-investors alike seek them out for the next great investment opportunity.

Structuring and Negotiating Venture Capital Investments

Most venture capital deals are structured as convertible preferred stock, the terms of which are negotiated by reasonable people who are all focused on creating value. The ideal way to structure an investment is in such a way that if the investors do what they say they are going to do (such as write the checks and provide strong support at the board level), and management does what it says it will do (such as hit its milestones on time and on budget), everybody owns equity that is valuable. If investors deliver what they promise and management does not, the structure should change so that their investment is protected. If management delivers and the investors do not, the investors should be diluted.

The single most common issue that we negotiate with entrepreneurs is governance—how will the board of directors be structured, and what happens if we decide that the company requires a new CEO? Secondly, depending upon the circumstance, we prefer to tranche our investments (providing funding in pre-negotiated chunks) so that the company knows that the committed funds are available, but the money goes into the company as it hits the milestones committed to by management. Tranched investments are important to us because, from the company's perspective, they know that the funding they need is committed and available as long as they stay on track. From the investor's perspective, it is possible to commit eighteen months worth of capital but maintain some degree of control if management is missing milestones. The elements that are most often negotiated in this case are whether or not to tranche the investment; the size and timing of the tranches; and the specific milestones. Investors want quantifiable value-creating milestones to be hit by specific dates (such as hitting a revenue target, successfully completing a clinical trial phase or closing a major collaboration with a partner).

Another factor that we as investors look for is a balanced board of directors, usually with representation of the common stock (management), but with significant representation by the investors and, in many cases, a tiebreaker agreed upon by both. There are many other terms that are specific to the circumstances of the deal and will change if you are dealing with earlier stage/later stage, performing company/struggling company, up round/down round, etc. No matter what the situation, balanced board representation is important because the board controls company governance, future fundraising, terms of sale, and, most importantly, decisions regarding management. If company management, and/or their representatives, control the board, it can be very difficult for investors to protect their investment through enforcing cost reductions, management changes, or sale of the company. Management often does not want to share control of the board with investors for the very same reasons.

Finally, a significant issue is the amount, form, and timing of employee/founder ownership of the company. Particularly when making the first investment in a company, how employees and founders participate in the equity ownership of the post-investment entity is critical. Prior to taking money from investors, the employees have often granted themselves stock in the company, and may own 100 percent of the company's equity outright. When we as investors write a large check, we too are purchasing equity in the company. Often, investors will require that the employees and/or founders change the form of their ownership from outright ownership of stock to stock options which "vest" over three to four years.

Vesting means that each year the employee works for the company, they acquire ownership of a proportional percentage of the stock options that have been granted to them. Should the employee leave during that vesting period, the unvested options will revert back to the company. Investors want employees to have a material equity stake in the company, because creating value in that equity will benefit both investors and employees. At the same time, investors are loathe to have employees who leave the company (and hence do not continue to create value for the entity) own a material amount of stock, as that stock is required to incent new employees to join the company and add value.

Often, founding employees believe that they are entitled to a disproportionate ownership of the company's equity because it was "their idea," or "they started it." Unfortunately, starting things is relatively easy, it's finishing them profitably (back to execution risk) that is very, very difficult. We as investors do not get our money back with a return until the job is finished profitably, and we want to make sure that the employees who are most highly rewarded are those who are instrumental in finishing, not starting, the job.

Every Deal is Different

There are hundreds of variables that factor into structuring a venture investment including but not limited to:

Valuation
Form of investment
Amount of investment
Terms of the investment vehicle
Anti-dilution provisions
Composition and structure of the board
Decision latitude of CEO
Reporting requirements of company
Employee stock amount/form/vesting
Distribution of profits at sale
Registration rights
Conversion ratios and prices
Future pro rata investment rights
Voting agreements
Rights of first refusal
Tag along and drag along rights
Right of co-sale
Warrant coverage and pricing
Dividend rates and payment terms
Reimbursement of directors' expenses
Amount and terms of D&O coverage

Protective provisions
No-shop clauses
Indemnification clauses
Company and investor warrantees
Who pays the legal fees
Notification requirements
Conditions precedent to close

Get two lawyers in a room, and they will (usually for an exorbitant fee) come up with a couple of dozen more. The later stage the company, the higher the valuation, and the larger the investment, invariably, the more complex the terms of the deal. The three things that matter the most, though, are how much money is going in, how much of the company it is buying and in what form, and what the governance of the company will be after the investment closes.

Fixing the Problems

Often, management wants investors' money but is reluctant to accept the governance that goes with it; once management agrees to accept an investment, they are agreeing to share control of the company. If the management team delivers what it has promised, then everyone will be focused on how to maximize the opportunity. The challenging issues occur when the company underperforms and the investors are dealing with problems—very often big problems—that require more money from investors to fix. If it is our money that will fix the problems, we must have the ability to make sure that the problems are addressed at what we determine to be their root cause. Relinquishing this control is one of the hardest things for an entrepreneur to do.

Adjusting the Deal Structure

Deal structures can be different depending on whether the company is a start-up versus a more established company. A more established company is one that is delivering on its plan and shows the promise of true value creation. When this is the case, valuations are higher, ownership percentages are lower for a given amount of money, and there is probably a happy group of investors already at the board table. But if this more established company is missing its plan, running out of cash and has an unhappy set of current investors, then the new money will often set the terms.

Similarly, if a "start-up" company is a first time entrepreneur with a couple of employees and an idea that needs to be proven, it is unlikely that we would be interested at any price (although others certainly might). Alternatively, if the "start-up" in question consists of a CEO who has created significant value in a previous company and has recruited a first-class team around an area in which they are experts, then the deal structure becomes negotiable. This is particularly true if the CEO can provide customer/partner prospects and strong references of past success, or if the CEO can create competition among venture firms!

There are not many factors that would lead us as investors to significantly alter the deal structure we offer to an entrepreneur. Situations that result in normal deal structures being thrown out the window are rare, but they usually are inspired by superstar CEOs or the surfacing of brilliant ideas. Factors such as whether or not a competing firm is trying to invest or if the potential investment shows spectacular results are always considerations, but what it boils down to is whether or not we truly like the prospect. If the prospect seems strong and we have good instincts about it, we will work with the management team to try and convince them that we are valuable partners. If we are not interested, we do not really care about the structure, price, or other investors. In fact, if we are tempted to change our mind based on those factors, we really should not be making the investment; this is a lesson we have learned over several years.

Establishing Valuations in the Biotech Industry

Establishing valuations is the essence of the venture investing art. The most successful biotech investors are almost binary in setting valuations, meaning that the valuation is either zero (a.k.a., "we're not interested") or "reasonable." Since valuations of private companies rarely jump tenfold absent a major milestone, "reasonable" is a negotiated term. Hindsight reveals that there is not that much difference in the ultimate value of a successful investment if the pre-money valuation varies by 10 or 15 percent in either direction. In an unsuccessful investment, there is a profound difference between having made the investment and having passed on it.

Valuations are ultimately based on the due diligence that is performed prior to making an investment. In biotech investing, we will always bring in outside experts to help us evaluate the science and intellectual property. We also do exhaustive reference calls on company management.

In our due diligence, we construct our own financial models that incorporate market size, penetration, future financings and the associated dilution, employee options, and all of the other standard fare. At the end of the day, though, the one thing you can be sure of is that all of the numbers are wrong; in venture capital, the future is truly unknowable.

For us, the first step in setting a value on a new company is to determine if we genuinely like the CEO and management team and have confidence in them. If we can answer yes, we can proceed to the next step, but if we are hesitant, the valuation is effectively zero (we pass). The second step is to ask if the product in question meets a large, unaddressed need with little or no significant competition. If the answer to this is yes, we can move on to step three; again, if the answer is no, our valuation is zero. The third step is to establish if we can get to a major value-creating milestone on a small amount (i.e., less than $5 million) of our money. If the answer is yes, we proceed to the next step; if it is not possible, the valuation is zero.

Finally, we determine if there are strong people associated with this company, such as co-investors, Scientific Advisory Board, and the Board of Directors. If not, we go back to steps one through three and figure out what we missed. Once satisfied, we go to step five, the final phase, which is to negotiate a fair price with the company. The most important variables when establishing valuations are, above all else, management, management, and management. Once we have established that management is solid then we can consider both market size and capital intensity.

The numbers on valuations for various types of scenarios, such as start-up companies, entrepreneurs with hot ideas, or companies on the verge of going public, can vary dramatically. There are some start-ups that might command a $50 million valuation and some public companies whose enterprise value would actually be a negative figure. This wide range again emphasizes the importance of factors mentioned throughout this chapter: intuition, judgment, luck, reputation, and results.

Summing It Up

Biotech venture investing is one of the most rewarding professions one could hope for—working with smart people on small teams focused on next generation technologies that have the ability to save lives and improve the quality of life, and at the same time offers the opportunity to create value for yourself, your partners, and employees of the company. Every day is a learning experience.

Biotech venture investing is also hard work and fraught with risk. Each new investment is a multi-year, multi-million dollar commitment that will invariably go through multiple ups and downs. Some investments that appear to be sure winners at the outset will end up losing everything. Other investments that look like they are lost causes will end up "fumbling the ball across the goal line" and making a lot of money, and for all of them, it's not over 'till it's over.

For many kinds of VC biotech deals, we can maximize the probability of success by focusing on building businesses, and adding as much value along the way as possible—from strategic direction to employee recruitment, from partner introductions to CEO coaching. In partnership with management, we will apply judgment and hard work, and hope to benefit from some luck along the way. And even with all of that, there's no guarantee that the business will be successful. That's why they call it "venture capital."
Duh!

For the past 28 years, Wm. Blake Winchell has focused on private equity investments with Leucadia National Corporation, the Channel Investment Group, Generation Ventures, and the Fremont Group. Mr. Winchell is a managing director of the Fremont Group, a $6 billion private investment company with investments in real estate, private equity, venture capital, and public equities. He is also the managing general partner of Fremont Ventures, the private equity growth-capital fund of the Fremont Group. Fremont Ventures typically makes investments of $2 to $20 million backing professional management teams with scalable, technology-based businesses. Fremont Ventures was *recently named by* EE Times *as one of the top ten venture capital firms in the U.S. for adding value to portfolio companies.*

Mr. Winchell graduated from Dartmouth College in 1975 as a Rufus Choate Scholar with a B.A. in English and economics where he was awarded the Henley Economics Prize. He received his M.B.A. from the Stanford University Graduate School of Business. He joined Bain & Company in 1978 and was one of the founders of the San Francisco Bain office.

Mr. Winchell is a certified public accountant. He has been a guest lecturer and featured speaker in the following forums: Haas School of Business, University of California, Berkeley; Financing Quantum Growth Conference–YPO; World Bank Entrepreneurship and Venture Capital Training Program; SEC Annual Government-Business Forum on Small Business Capital Formation and Cross-Border Investing; Capital Access Conference on Value Added Investing; and Cambridge Healthtech TriGenome Conference.

Mr. Winchell is chairman of the Board of GlobalServe Inc. and a board director of BioSeek Inc. LightSand Communications, and eSilicon. Blake is on the Advisory Board of the J. David Gladstone Institutes and serves as the Chairman of its Steering Committee. He is also a member of the executive board of the Skyline Council, Boy Scouts of America, World President's Organization, and serves on the International Advisory Board of the Instituto de Empresa in Madrid, Spain.

Keys to Success: A Vision for Venture Capital in the Biotech Industry

Heidi L. Huntsman

Partner

UV Partners

As with all industries, investing in biotech is cyclical. Unlike most other industries, biotech companies have a very long regulatory pathway, and it can take years (seven to ten) in product development before knowing if your therapeutic is efficacious, let alone realizing any revenue. Frequently large amounts of capital is spent before there is an indication that the drug in development will have efficacy in humans, even though it seems to cure all sorts of diseases in rodents and other animal models. The capital requirement for human clinical trials and manufacturing scale-up of clinical grade drugs is immense.

In past years, a biotech company could ride high on an IPO based solely on the "promise" of a drug in clinical trial. Unfortunately, investors have been burned time and time again with promising drugs failing in the final phase of clinical trials. Today, savvy public investors need to see more clinical data before placing a high valuation on a promising biotech company; therefore many companies are having to postpone going public, increasing the capital burden for the private investor and extending the horizon for an exit by means of a public offering.

From 1999 to 2005, the average amount of venture capital raised by a start-up biotech company before an initial public offering (IPO) has remained pretty constant, hovering around $100 million. However, the average return at IPO has dropped from 5.3 times in 1999 down to 2.3 times in 2005. In addition, the time from first venture investment to IPO has increased from 5.4 years in 1999 to 6.3 years in 2005. Keep in mind these returns are generally binary—the averages are the result of some high returns, coupled with some breakeven and some losses, and with the low number of companies in the pool, it's hard to get a true average. However, the general trend is clear—high returns in this sector are less frequent, and the time to IPO has been increased. This environment makes investing in early stage biotechnology companies today very difficult, because you can't count on a public offering as a way to get venture returns. In addition, the IPO market for biotech companies has been soft, and companies have not been holding their value on the public markets. Often by the time a venture investor is able to sell its stock on the public market (generally the stock is locked up for six months following an IPO), the investment is either under water or relatively flat. Considering the risk involved, the time elapsed, and the amount of capital required prior to an exit, many firms have been rethinking their biotech investment strategy. In fact, some funds are getting out of the space altogether. Angel investors and smaller funds that typically do seed and early stage investing have been getting squeezed in later rounds, and they do not have the capital to follow the investment through. Large funds can place big bets on later stage companies, where much of the technology risk has been taken out and the road to liquidity is fairly clear, or at least less risky. Yet even in these cases, returns are not what one would expect from an industry with so many potential blockbusters. Given these challenges, the question is: what about the small start-up? Where do they find capital and what can they do to keep things moving forward while they are raising it? There are some basic tenets and tactics that will strongly increase the likelihood of success for such companies.

An essential element to venture capital success in the biotech industry is maintaining focus. While a pipeline of potential drug candidates, and/or a platform with which to generate proprietary candidates is important, it is crucial for a small, resource-constrained company to keep the focus on developing one drug at a time until there is evidence of safety and efficacy in humans. Once the platform has been validated and evidence of human efficacy has been established, capital should become available through an industry partner, additional venture financing, or even possibly an IPO, which can be used to move subsequent pipeline products though human clinical trials. Additionally, venture capitalists do not like to fund research projects. If development of your pipeline involves more "R" and less "D", it is important to be creative: small start-ups should identify non-dilutive funding sources to keep the basic science moving forward. Examples of such funding include NIH grants, small business grants such as SBIR and STTR grants (http://grants.nih.gov/grants/funding/sbir.htm), foundation support, and many other alternative routes.

Another element of success is investing in a strong management team. Before making a significant investment, it is important to identify teams that have successfully managed a similar venture before and are experienced with the process of taking drugs through clinical trials and to market. Do not take a risk with an executive or team who may make novice mistakes on your dime; although many delays in development are out of the control of the management team and may happen regardless of how experienced they are, setbacks through novice mistakes can significantly push back progress.

To increase the chances of success, fill out a skeleton management team with top-notch experienced consultants until you know enough about your novel compound to build a company around it. Hire the best and most experienced consultants available. This allows management to have more control over the burn rate by increasing or decreasing the amount of time needed from consultants. Generally, small companies cannot afford to hire full-time employees of this caliber early on, so a team of consultants can bring great minds to the development of the products without committing to a high monthly burn.

Making Successful Investments in the Biotech Industry

As discussed above, recruiting excellent management is a crucial first step in a successful biotech endeavor. Although very early stage companies will not have the resources to fill out the entire senior management team, make sure there is at least one seasoned biotech executive who is a dedicated full-time employee, with ample motivation to move things forward.

Another critical component of making successful investments involves understanding the clinical landscape. Establish that a clear unmet need exists and will be addressed with the therapeutic of interest (i.e., doctors care about this drug and will likely prescribe it to their patients because right now they have few or no options). Understand why other drugs in this space have failed (mechanism of action not verified in disease, toxicity, failed to show efficacy, etc.). Further, investigate the disadvantages of current therapies and determine what is different about the mechanism of the compound or product in which you are investing. Be sure the potential advantages are clear and scientifically or clinically validated. With respect to support and evidence, research how closely the animal models mimic the disease of interest. Find out as much information as possible regarding the structure of the clinical trial, including factors such as size, ease of patient recruitment, clarity/clinical acceptance of trial endpoints, time to follow-up, and the cost involved.

Intellectual property (IP) is a crucial component in biotech investing. Understand first who owns the IP, and what rights the company has to it. Read the contracts, understand what has been filed, what has been issued, when and where (which countries), and do a thorough review with the company's IP attorney for starters. IP is such a critical barrier to entry in this industry that once we make a decision to put out a term sheet, our investment is contingent upon a thorough IP review conducted by an outside law firm that has expertise in the space to verify that the company has freedom to operate (FTO), as well as protectable IP. This is an area where early investment is important—find a very experienced IP attorney who can be strategic around filing patent applications and can build a picket fence around the core IP to block it from every angle imaginable. This is the only true asset of an early stage biotech company and must be rigorously protected. The valuation placed at the exit, whether an acquisition or an IPO, can vary several fold depending on the strength and breadth of the IP.

In general, it is fair to say that it is better to be lucky than smart. Animal disease models are not always predictive and human diseases are typically multi-genic—thus, luck can be the number one factor in success in this industry, especially when investing in preclinical therapeutics. No matter how solid the animal data, it is not possible to know with certainty how safe and efficacious the drug will be in humans until it has been through human clinical trials. Additionally, with cyclical markets, you can never predict what the market environment will be like when you are ready for an exit. Therefore, it is best not to get too tied up in trends and hype; just because it's a hot market today and others are overpaying for it, doesn't mean it will remain a hot market leading to high returns upon an exit in the unforeseeable future. Look for solid science, excellent management and leadership, good value, and reasonable goals/time frames. Finally, because luck and timing are generally not factors you can control, it is important to place several bets. Past industry analysis has shown that a good rule of thumb is that ten investments with great potential will generally yield one investment with a great return, three with good returns, three that break even, and three that get written down or even off.

Unique Approach to Biotech Venture Capital

Our venture fund is relatively small, and we are only able to invest about $8 to 10 million total over the life of a deal, so we have developed a unique approach to investing in biotech. Because we are small investors, it is imperative that we think long and hard about how we will be able to continue to support our position. In many ways, this is positive, as it compels our evaluation process to be both thoughtful and thorough. Generally, this approach means getting involved early, sometimes even while the drug is still within an academic institution. It also inspires us to structure the deal creatively to get further out of a small investment and to ensure that we have other deep-pocketed funds in the syndicate or teed up or the next round, but preferably as part of the syndicate.

As part of our investment approach, we strive to spot opportunity in the industry through aggressive networking and clinically focused conferences. Our approach is to find an area of interest, learn as much as we can, and understand what the biggest needs are. We then identify a start-up in the space with proprietary technology, strong management, great scientific minds, and solid data.

In structuring a biotech investment, we like to tranch our investment with very clear and specific milestones in order to mitigate risk. For instance, if we raise (with a syndicate) $15 million in a series A, we are committing to $15 million, but only investing a portion at a time: it may be three tranches of $5 million each, or some other combination structured to achieve certain milestones along the way. We hold the management accountable for achieving those milestones before the next tranch of money goes in. It is amazing how quickly and efficiently things get done if the issue of making payroll is at stake. This accountability is not meant to be a tool for micromanagement, but rather a means of providing a clear focus in the critical stages of early development. It will hopefully help align the interests of all parties. Of course, not everything will go smoothly: things cost more than you planned and the process always takes longer than anticipated. Given these challenges, we work collectively to figure out how to achieve the milestones on the resources allotted or come up with alternative solutions. We aim to be supportive of a management team that is working hard, even when unanticipated hurdles arise. If there are obvious signs that the project should be aborted, it's crucial to maintain clear and open communication with the management and work together through the process. By tranching the investment, we are less at risk for blowing through the entire commitment before becoming aware of an issue, instead of evaluating each incremental milestone along the way.

A model we like to use for cutting down on the immense capital requirements and to minimize burn rate is to be "virtual," by which I mean the company keeps its science in an academic setting—assuming that is where it started—for as long as possible. The capacity to remain virtual will obviously depend on the institution, but can greatly reduce overhead cost and capital equipment purchases, while bringing more scientific resources to the project. If doing the work in an academic setting, it is extremely important to negotiate an up-front deal with the institution for first rights to any related new inventions, establishing the same terms on all inventions, and keeping the royalties low. If done wisely, this can be a win-win scenario for all parties; a start-up can save a significant amount in development costs while keeping the best minds engaged in the development of the technology. An academic institution can benefit by (eventually) having an incoming royalty stream from ownership of the IP, as well as recognition for inventing a blockbuster drug. Additionally, academic institutions that support this type of structure are less likely to lose their best scientific minds to industry, because they allow them to peruse both.

The Greatest Challenges and Mistakes with Respect to Biotech Venture Capital

With respect to the many challenges in biotech venture investing, the regulatory process is probably the most brutal. It never seems to be black or white, just different shades of gray. The shade of gray depends on many factors, such as the amount and type of information already known about the safety of the compound category, the current environment (specifically, have there been recent safety issues that have resulted in negative publicity for the FDA), and certain pet peeves of the individuals sitting on the FDA review panel. Setbacks with the FDA can result in months of delays, and you can essentially count on this type of setback happening at least once during the process. Much of this is the unfortunate result of the liability the FDA has with approving trials administrating novel, potentially harmful drugs to human beings combined with constrained resources. The FDA has to respond to an investigational new drug (IND) application within thirty days. If your FDA committee is backlogged and has not been able to fully digest your 4,000 page document within the thirty day time frame, they will almost certainly find some reasons to put your trial on clinical hold in order to buy an additional forty-five plus days for a more thorough review (turnaround time in correspondence combined with a new thirty day time frame for response). The best-case scenario one can hope for at this point is that the additional information requested is easily obtainable and does not require further animal or toxicology studies. This process is always on the critical path and very difficult to navigate; for these reasons, this is an area for which it is essential to hire highly experienced regulatory consultants to help you through the process and navigate the FDA.

In the FDA's defense, they are aware of some of these hurdles and the impact they impose on capital constrained start-up biotech companies, and therefore on the pace and creativity of new drug development. However, their absolute primary concern is to ensure, as best they can, that novel drugs going into humans will not cause harm. Although resource constrained, they have made efforts over the past couple of years to restructure the process to make it (ever so slightly) less onerous, without compromising safety. An example of this is their new critical path initiative, which is being established to renovate the process of drug development by modernizing the science behind how they go about evaluating new molecules. This could take weeks or even months out of the regulatory process, but will most likely be years before it is fully implemented.

The most frequent mistake venture capitalists make is underestimating the amount of time and money it will take to get to an exit or a critical value inflection milestone, and they don't reserve enough "dry powder" to see the investment through. The company ends up being chronically undercapitalized and the problems get exacerbated as the investors become "tired." There are so many unpredictable delays in drug development, including the ability to scale up manufacturing, dealing with the FDA, and slow patient recruitment in trials, that it always takes much more money and more time than expected to get an exit. Another common mistake is waiting too long before making critical changes—either to management, product focus, manufacturing or development. Generally when something begins to go wrong, the writing is on the wall long before change is implemented. However, when you've invested a lot of time/money/energy in a strategy, it's hard to be sure when to call it quits and human nature is such that you would like to make it work. Additionally, different co-investors or board directors may have different opinions, and instituting change may take some time and persuading of fellow board members. These challenges and delays are constant across the board in most early stage companies but are particularly prevalent when investing in biotech due to the long time horizons, capital requirements, and lack of certainty before knowing that you have a product you can take to market.

Growth Potential: What to Look for in a Company in Which You Are Investing

Because it takes so much capital to take a drug through the drug development process, as investors, we generally only consider investment in a drug that has a market potential of >$500 million, preferably >$1 billion. When investing in an area where there is very clear clinical need, a successful drug has the potential to ramp very rapidly if it has been partnered with a company with an established sales force. For example, Tysabri, a Biogen-Idec drug to treat multiple sclerosis (high unmet need with very defined patient population and established sales force), was anticipated by many to ramp to roughly $1 billion in revenue in its first year on the market. Unfortunately, it was pulled off the market after three months due to adverse events that lead to some fatalities and Biogen lost roughly $11 billion in market cap, with the stock falling from $66/share to $34/share. Because of the extreme unmet need in this particular disease, Tysabri was recently allowed back on the market with black box warnings (FDA imposed restrictions and warnings that must be present on the label). The stock has only rebounded modestly, and has been hovering in the mid $40s. Although this case is extreme, these are the kinds of highs and lows one can anticipate when investing in biotech—it is not for the faint of heart.

Assuming the drug is able to show efficacy, and is a much needed improvement on current therapies, I would argue that the two essential components to rapid market penetration are: a very clear unmet need with a defined set of physicians who understand the importance this therapy will make in the health and well-being of their patients, and a strong patient advocacy group that will get behind the therapy and promote it from within (as with the case of Tysabri). Outside of that, strong partnerships with companies that have marketing muscle and an established sales force are key to rapid market penetration.

Becoming "Successful" Over the Long Term

A definition of success, as you may imagine, comes down to making more money than you lose—hopefully a lot more. Although stated simply, if it really were simple, everyone would be doing it. No matter how well things are going, the extent of success of an investment is only measurable upon exit. Developing good partnerships with large pharmaceutical players is a great start, and can provide market validation, cash in the form of milestone payments, and large company resources, but until the venture investor can get its equity out of the company, you don't know how big the return will be, or even if it will be successful; so many unanticipated setbacks can occur at any point in development, and even after commercialization (e.g., Vioxx). In the current investment environment, even most IPOs are not looked upon as exits; rather, they are viewed as additional financing events, and venture investors are commonly under water in their investment upon IPO. Thus, even though the equity is technically liquid after an IPO has come out of lockup, often the investor needs to hang on to its position for several months, possibly even years, after an IPO to see the increase in valuation that comes with a successful drug approval and launch, or subsequent acquisition.

There is currently a lot of merger and acquisition (M&A) activity by large pharmaceutical companies, which can result in a good return on investment, and has the potential to be an earlier or at least more viable option than taking the company public. Large pharma is struggling from a lack of promising drugs in the pipeline, and most tend to be building their pipelines through acquisition. However, if the drug is in a very early stage of its development, the acquisitions tend to be back-end loaded: to receive the full buyout price, the company has to hit milestones established by the acquirer. Therefore, even in the case of an acquisition, the investor still may not know exactly how successful the exit will be for many more months, potentially even years, until the company has been able to hit the milestones required for the full payout amount. In general, the further along the drug is in development, the higher the up-front payment.

The best chance of increasing success upon an exit, especially if you are considering acquisition, is to ensure you have several suitors for your technology. Start early in developing relationships with potential partners.

By having developed a relationship early, when it comes time to discuss acquisition, the interested parties are already familiar with the potential of the drugs in development, and have a degree of comfort that comes with following technology development over time. Parse up your technology into packets that allow you to talk to as many potential partners as possible—i.e., one group of suitors for development and distribution rights of drug one in Japan only and one for distribution rights for drug two in Europe and Japan, etc., keeping the "jewel" (generally drug one in the U.S. market) free to either pursue internally or to entice an acquirer in an acquisition (i.e., the only way they can get that particular drug in the U.S. market is to acquire it). By setting up these kinds of relationships, you will likely receive milestone payments, up-front license fees, and sometimes even have your development costs covered, depending on the partnership, all of which can help move the core technology forward and validate the therapeutic potential.

The Most Common Structuring of Venture Capital Deals

Structuring venture capital term sheets is very dependent on market environment. Generally, if the market for exits (IPOs or M&A) is "hot," deals are more competitive and valuation and deal terms tend to be more entrepreneur friendly. When liquidity options are scarce, it becomes much harder to raise venture money and terms favor the venture capitalists. In today's environment, for a company that has a mature management team and has been hitting its milestones, most deal structures tend to be pretty standard: participating preferred with a one times preference. What this means is that upon an exit, the preferred equity holders receive the amount of their investment back in preference to any other equity holders. The remaining proceeds then get divvied up on a "pro-rata" basis among all the equity holders.

Today, preferences are generally around one times, and participation is capped around two to three times the amount of the investment, although these parameters vary greatly based on market conditions. For example, in the bubble of 1999 to 2000, the environment was much more entrepreneur friendly and preference and participation tended to be capped at one times investment (if there was a preference at all). After the bubble burst, the table was turned in favor of the investor, and preference and/or participation could be as high as five to six times the investment. The amount of preference can range from no preferences (all preferred stock participates pro rata with common), to as high as six times in a really tight market (where investors with preferred stock are paid out six times their investment before the common shareholders see any returns). Another term seen frequently in a bad market is a full ratchet. This can protect an investor from a future down round by increasing the number of preferred shares to effectively lower the price per share to the price of the new (down) round. Anti-dilution terms are similar, but generally protect the investors percent equity if additional shares are added within a certain time period (i.e., if additional option or warrants are added after the current financing round has closed that drop the investors equity from 10 percent to 9 percent, the investors will be allocated additional preferred stock to bring them back up to 10 percent).

To clarify some of these terms, let me provide a simplified and straightforward example of one times participating preferred, capped at three times participation: If Investor 1 invests $10 million in Company XYZ and the post money valuation is $100 million on a fully diluted basis (this includes an option pool for management), then Investor 1 owns 10 percent equity post-financing (Table 1). Investor 2 had invested $20 million in Company XYZ in a prior round and is junior to Investor 1, but also has a one times preference. If the company gets acquired for $80 million in cash, Investor 1 will receive its full investment back ($10 million) in *preference* to any other equity holder. After Investor 1 gets its preference, then any junior preferences (in this case Investor 2) will be paid out ($20 million) before the common shareholders see any return. After all preferences have been paid out ($30 million in this example), the remaining $50 million will be divvied up on a "pro rata" basis among all the equity holders. So Investor 1 would receive 10 percent of $50 million ($5 million) for a total payout of $15 million, Investor 2 would receive $20 million in preferences and $10 million from participating ($30 million total), and the common shareholders would end up with 70 percent or $35 million. This is referred to as participating preferred. In this case, the cap on participation is three times, so in cases where returns are higher than three times, the investor would choose to convert its preferred shares into common shares, and participate pro rata with all common shareholders. For example, in the scenario of a $400 million acquisition, the investors would be worse off if they kept their one times preference (Investor 1 would be capped at $30 million due to the three times participation cap, where if they converted, they would have returns of $40 million). This structure works to protect the investor's investment in the case of an underwater exit (they get their full investment back before the common holders), but also serves to provide motivation for the management to push hard for a return greater than three times (or whatever the cap may be).

Goals of VC Deal Structuring

From a venture capitalist's standpoint, it is critical to give the company in which you are investing ample runway to solidly achieve a value inflection point, while keeping the post-money valuation below the level at which you anticipate the next round to occur. This will vary depending on where the company is in development, but is generally pretty straightforward. For example, the goal could be achievement of a milestone such as completion of Phase I trials or the signing of a large partnership agreement. The time and capital requirements to reach these achievements can be modeled and additional cushion should be added for unanticipated setbacks. If the company is undercapitalized and not able to reach milestones that are critical for an increase in valuation, the investors will not likely see a step-up in their investment until the company is able to reach those milestones. This could result in a down round (where the new investors come in at a lower price than the prior investors), or added runway via a bridge loan from the existing investors to tide the company over until they can reach the intended milestones. Drawing out this process due to chronic under-funding can result in an unnecessary outflow of cash due to extension of the burn while waiting to be capitalized.

If the company ends up in a situation where it needs a runway to get it through a rough spot, the current investors ("insiders") will frequently provide a bridge loan to the company. This loan can have a broad range of terms and conditions attached, depending on the state of the company. If things are going very well, and the company just needs more funding to hit important milestones before seeking additional venture financing, the terms will most likely be pretty gentle, for example: 10 percent warrant coverage, 6 percent interest, and conversion into equity at the next financing round (again, these terms will vary based on market environment and potential of the company). If the company has consistently missed milestones and is in trouble, the bridge will likely be designed to get them to an immediate exit (quick sale most likely) or extended runway to get things turned around and back on track. Bridges under these conditions generally have much more onerous terms, and the bridge loan can be structured so the investor has the potential to get the bulk of its original investment back (for example, a five times preference on a bridge loan in the case of a quick sale or heavy warrant coverage if the expectation is to do an additional financing at a down round).

As discussed earlier in this chapter, we like to tranch our deals according to certain milestones that we would like to achieve during the financing period, so we tend to make this process even more granular, breaking it out on a milestone by milestone basis.

Considerations for Venture Capitalists When Structuring a Deal with a New Company

The most often negotiated items with the entrepreneur tend to be pre-money, the size of the option pool, and preferences. In terms of structuring a deal with a company in which you may potentially invest, valuation is a critical factor for venture capitalists. There is generally a range that one would consider, but this range is pretty narrow unless you are considering a highly competitive deal. As a small firm, we will not play that game, because we have to get in the investment at a good value in order to maintain our equity position over the life of the deal—i.e., we are valuation sensitive. If we cannot come to terms on valuation, it becomes pretty easy to walk away from the deal. Another important item to consider is the option pool, which falls under valuation, as it directly affects it. There needs to be a big enough option pool factored in to the post-money to get through the financing period, otherwise you end up getting diluted by additional options (assuming they all eventually vest).

The amount of capital raised is also important. It is critical to make sure the company has ample money to reach its intended milestones so you do not end up having to raise additional capital before you have hit your value inflection point. A less seasoned entrepreneur may want to keep the amount of capital raised small so that he or she will not get as diluted. This approach can be shortsighted, as being under-capitalized can lead to greater dilution down the road when the company runs into trouble or is unable to reach the milestones needed for a bump-up in valuation.

Another important factor with respect to the terms of the deal is the voting structure. From a venture capitalist's standpoint, it is necessary to structure the term sheet so that any one party does not have the ability to block a decision or potential transaction. Such deals include a partnership opportunity, acquisition, new investment, or other opportunity in which you might be interested down the road.

Board composition is also very important when structuring a deal. Make sure your co-investors and management team's interests are aligned with your own. If they are looking for a quick "flip" and you are more interested in developing a stand-alone company, you may be in for trouble down the road when it comes to investing in building infrastructure, or with others wanting to take an early acquisition offer that you believe leaves a lot of money on the table. It's very important to have these conversations before getting in bed together, so that expectations for the future of the company are well defined and a unified strategy is understood by all parties. Things are always subject to change, and individuals need to remain flexible, but it's important to know you're on the same page to begin with.

Make sure the personalities and experience base on the board is diverse, but the dynamics are good: you want to avoid assembling a bunch of people with a similar background representing the same opinion. Critics can be very beneficial, so long as they are constructive and not continually causing waves. Do due diligence on the individuals who will be on the board, just as you would with the management team. If they are not individuals who you've worked with before, it's important to find out what they are like to work with on a board so you won't get too many surprises. Also, it is wise to have at least one outside industry expert on the board. Such an individual can add tremendous strategic value and, hopefully, will come with an industry rolodex of contacts and information sources.

Misguided Interpretations: Trouble Spots for Entrepreneurs

Entrepreneurs are most misguided when it comes to valuation expectations. Entrepreneurs tend to believe they have the greatest drug and strongly feel it should be highly valued because it will go through clinical trials without any hitches (of course it's safe!), hit the market in record time (assuming it's not snapped up in an acquisition first), and achieve its full potential within the first couple of years on the market. This optimism is understandable, even somewhat desirable, because you want the entrepreneur to be passionate about pushing the technology forward. But you also need the entrepreneur to be realistic and practical, and in touch with the market. They need to understand the value of what a good biotech investor can bring to the table in addition to capital: contacts, strategic insight, operational experience, and access to additional capital—all which serve to move the company forward, and create something that is a much larger opportunity in the long run.

One challenge that we frequently encounter with early stage companies is that the company has been funded through grants and unsophisticated angel investors; as a result, it has ended up with an unrealistic valuation, from which the involved executives expect a step-up because they have made so much progress (bottom-up analysis). Unfortunately, they are usually out of touch with the reality of venture valuations, which are a result of years of experience in understanding how much money it takes to bring a drug to market (top down analysis). By being firm about an unrealistic valuation, the entrepreneur is doing him or herself, as well as the company, a disservice because the company can become tainted. Once it is known that several VCs have looked at the deal and walked away, it becomes very difficult to attract new VCs that are willing to take a look; unfortunately, this is a small industry and most people talk. A negative reputation, once earned, is easily spread and difficult to shake.

Because of the disconnect between how an entrepreneur sees the valuation of his or her company and how the VC sees the valuation, there can be a lack of trust in the motives of the venture capitalist. What the entrepreneur needs to keep in mind is that the venture capitalist's primary motivation is to make money for his or her investors. If he or she is able to accomplish that goal, the entrepreneur ends up making money as well. With the additional capital and professional strategic insight that comes along with seasoned venture capitalists, the potential to grow the company into something very large is much greater than if the entrepreneur wants to boot strap and plug along on his own. Thus, although the entrepreneur ends up owning a smaller piece of the company, that small piece of equity generally ends up being worth much more than the whole piece would have been without the additional resources. Venture investors usually will not be interested if they cannot make several times their investment.

Factors that Can Alter the Deal Structure Offered to an Entrepreneur

There are some factors that can alter a deal structure that we will potentially offer to an entrepreneur. As the syndicate comes together, generally the new investors will want to add in some of their own terms. If they are a significant investor, there is usually room to negotiate new terms. When debating deal structures, room for negotiation varies greatly from firm to firm. Generally, the more competitive the deal, the more room for negotiation. Also, the rapport that the entrepreneur has developed with the investor can come into play in negotiations. If the company has been looking for financing for a long time and has no other options, it most likely will not have a lot of negotiating power.

One scenario in which we would be willing to make alterations in deal terms (prior to close) is if the company hits a significant milestone during the negotiating process, such as a new partnership opportunity with an up-front milestone payment. If this is the case, then the fundamentals of the deal have changed, and thus, there should be some negotiation room. This can go in the other direction as well—if the company had anticipated hitting a milestone and doesn't during the expected time horizon, and achievement of the milestone has been baked into the deal terms, again, the fundamentals have changed, and there should be flexibility to negotiate down, or in some scenarios, walk away.

Given the variability of the many scenarios in which one may be investing and the unpredictable nature of the industry, it is critical that those investing in this field are realistic and able to consider the true value of investments. Yet, there is something to be said for intuition and passion as well; it is balancing these elements that is the trickiest yet most rewarding component of venture capitalist investment in the biotechnology field.

Method for Establishing Valuations in the Biotech Industry

Setting a value on an early stage company is more of an art than a science. With that said, there are certain hard facts that should be investigated in order for you to establish your valuation. One factor that must be ascertained is the current financing environment. At a basic level, if you are financing a company to a certain milestone, such as completion of Phase I trial, it is critical to know what the range is for valuation for a company at that stage. Make sure your post-money is well below the range, including a healthy option pool and six to twelve months of runway for the unexpected.

Another method is to start with a moderate to low scenario of a realistic exit and consider your minimal return expectations (top-down approach). Work backwards to determine the time and money needed to reach these milestones and establish your valuation from there.

In terms of specific financial tools or ratios that are helpful in the determination of valuations, market intelligence is critical. Specifically, it is helpful to know what the valuations are for other similar deals that are currently being made; such data can be gathered through networking. This information is a vital component both for understanding the state of the industry and in determining what valuation you might expect at the next financing. IRR analysis is pretty hypothetical for early stage companies, but we generally run several scenarios just to make sure we fall somewhere within the bottom to middle of a very broad range of possibilities.

Valuing a Company in the Biotech Industry: A Step-by-Step Process

The first step in valuing a company in biotech is to understand the market and where the particular therapy or therapies fall within the market. As part of this early phase of valuation, it is critical to determine the unmet need that this product addresses. Is this something physicians see as an issue, and does it address a clear point of pain? What is the anticipated physician adoption? This step of due diligence is critical to understanding the potential of the drug or platform. It requires several calls and interviews with leading physicians and scientists in the target area. Other factors to consider are the size of the market and the market potential. Further, you will want to understand how the market potential is broken out. Questions to consider are: What are the competitive drugs and what is the revenue range for these drugs? Another key component to this step of valuation is identifying the downsides of existing therapies or treatments, such as their side effects, efficacy, method of delivery, and expense to manufacture, and then closely considering the purported advantages of the proposed new therapy.

The second step in valuing a company in this industry is determining the strength of the intellectual property the company is offering. IP is a tricky area, and an analysis should always be done by an outside IP attorney who is familiar with the area. An IP analysis is an expensive proposition and is generally done after a term sheet has been signed. However, you should do enough legwork before extending a term sheet to know that the company most likely has freedom to operate (FTO), and whether the IP is protectable. This stage of research as it relates to IP will include several conversations with the attorney who filed the IP and keyword searches on www.uspto.gov, the U.S. patent database. You should also engage in as many conversations as possible with others in the field who may be able to shed light on existing IP that potentially could block the IP of the company for which you are establishing valuation.

The third step is ascertaining the strength of management. During this phase, the track record of the management team should be closely examined. Consider their ability to raise additional capital and the contacts they have in the industry. At this point in the process, reference checks are critical and should be done by evaluating provided references as well as some that were solicited. References not voluntarily provided by the management executives themselves often reveal important details about their potential.

The fourth step in establishing valuation is scientific validation. To determine this, it is important to examine peer-reviewed publications discussing the scientific basis of the mechanism of action, evaluating the data from multiple animal models and well understood/defined/proven target and multiple potential indications. Understand the bioavailability—has the company shown the drug can be delivered via the delivery method specified? For example, is it soluble? Can it be digested via the gut and still reach its target in an active form? Can the drug reach anticipated therapeutic levels using nanomolar quantities? Generally, much of this part of the diligence is hired out to an industry or scientific expert, as technical expertise is needed for thorough evaluation.

The fifth step is evaluating manufacturing and delivery. You will want to question how difficult and expensive it is to manufacture this product. Typically, biologics are much more expensive to manufacture than are small molecules. With biologics, sometimes the larger the protein, the more expensive it is to manufacture and the more difficult it is to work with. Further, find out if manufacturing can be efficiently scaled. Another important factor to consider is the company's experience in manufacturing. Who do they have on board to manage manufacturing?
Other questions to be asked are What is the delivery route of the therapy? Some examples of delivery routes are oral, IV, Sub Q, trans-dermal, etc. When looking at an indication where existing therapies are drugs taken orally and the company you're investigating is suggesting developing a drug with IV delivery which requires the patient to go to the doctor's office and have an IV put in for an hour every week, you might think twice about what the adoption rate of the new drug would be. However, if the disease is severe and there are no good therapeutic alternatives, the patient may not mind getting a bimonthly IV injection in order to mitigate the symptoms of the disease.

The Most Important Variables When Establishing Valuations

The most important variable when establishing valuations is the stage of the company. This is analyzed by categorizing the company as preclinical, Phase I, Phase II, etc. Generally, most emphasis is placed on the drug that is the furthest along in development or the drug that the funding syndicate considers to have the most potential. While these are not necessarily the same, the valuation will be placed largely on the therapeutic the funding syndicate thinks is most promising to take forward.

The second most important variables are the power of the discovery platform and strength of the potential therapeutic pipeline. Investors must consider if they are investing in one drug or in the ability to bring several therapies to market following the first signs of success.

With respect to valuation, there is really no standard recipe because there are so many variables; variations exist among the drugs themselves, as well as among the various pipelines and platforms. Another variable is the need for a new therapy—reimbursement will vary, and so too will many other factors that are critical to the valuation. There are certain guidelines, however. A start-up company with an experienced and well-networked CEO, strong basic science, a clear IP, and high unmet need is going to be at the higher end of the range, especially if it also has large market potential, established reimbursement, platform for future proprietary drugs, and Phase I safety data. A company that has a reasonable CEO with one or two in-licensed drugs for a market where there are already many drugs with no distinguishing features would be at the very low end of the range and, ultimately, probably not worth pursuing. Obviously most companies fall somewhere in the middle. If we are faced with an entrepreneur who has an idea we would like to build a company around, the valuation would be in the low single digits. A company within twelve months of filing an IND would receive a valuation in the low- to mid-single digits, whereas a company with a drug in Phase I would be placed at high single digits to low double ($8 to $12 million including new option pool); this valuation could be significantly higher if there is a promising pipeline targeting large markets. A company with a drug in Phase II could have a pre-money from high single digits to mid- to high-double digits. These different scenarios reflect the variability of valuations and the many factors that must be considered.

Like any other industry, placing a valuation on an early biotech company is not a straightforward process. The answer to what a valuation should be, of course, is that it all depends. The constants that are key factors to a valuation include management, IP, pipeline/platform, and animal model data. Beyond that, there are many different factors that determine a valuation, all of which are highly dependent on the specifics of the given scenario.

Heidi Huntsman joined UV Partners in February 2004 as a life science partner. Before joining UV Partners, Ms. Huntsman was director of research at Tenex Greenhouse Ventures, a venture capital fund focused on early stage biotechnology and medical device companies. In addition to her role as an investment professional, she was active in the management of several portfolio companies. Prior to working at Tenex Greenhouse, Ms. Huntsman worked in investment banking as an associate equity analyst in the medical device sector at NationsBanc Montgomery Securities (now Banc of America Securities). She also managed marketing, business development, and sales operations for San Francisco-based medical supply start-up MedChannel. Ms. Huntsman has ten years of experience in academic research and holds an M.S. in human molecular biology and genetics and a B.S. in physiological psychology from the University of Utah.

Strategies for Venture Capital Financing

Thomas L. Churchwell

Managing Partner

ARCH Development Partners LLC

Geographical Concerns

One of the many myths about venture financing is that, like other forms of capital, it all rolls downhill. In fact, at very early stages of company development, venture capital is very geographically parochial. Typically, VCs will not invest in seed or very early stage deals that are more than an hour's travel time from their offices. In their formative stages, deals are very volatile and require a great deal of hands-on attention, so it is impractical and imprudent to invest at long distance.

Another myth is that venture capital in the biotech industry is nationally homogeneous. In fact, venture capital is quite asymmetrically distributed with the vast majority of early stage capital situated within a few miles of the two coasts. When a start-up company has achieved some traction (variously defined but typically comprising a strong, experienced management team and a product that has proof of concept in a well-defined market [a B or C round in venture parlance]), venture capital will be available from a variety of sources. But as we have already seen, earlier stage deals will be dependant on local financing by angels or very early stage VCs.

Thus, entrepreneurs in regions that are underserved by more traditional forms of financing have had to innovate in their financing strategies as well as their product offerings. The next few paragraphs explore some of those alternative routes. Since our firm invests primarily in the Midwest, we will discuss from our experience, but most of the strategies will be applicable in any region not adequately populated with venture and angel money.

Capitalizing on Grant Funding

One of the most effective sources of early stage funding is grant money from federal, state, and local public and private sources. Used correctly, these funds can be very efficient and effective sources of funding. Since grants typically are not repayable and do not give the grantor any equity participation in the company, they are termed non-dilutive (i.e., they do not dilute the equity interests of other shareholders).

One very popular source of such funding is the SBIR (Small Business Innovation Research) program operated by most funding agencies of the federal government. These are usually obtained by responding to specific RFPs (requests for proposals) published by the various agencies. The art to using SBIR grants is to avoid accepting grants that will cause the company to wander off its strategic pathway. Because the funding is limited in amount and scope, use of SBIR grants inherently requires efficient allocation of funds. But many companies have used SBIRs to successfully traverse the early stage funding chasm. Lacking payback requirements, SBIR funding can be used to enhance a company's value proposition without diluting shareholders. There are many similar state and local programs that either bridge to an SBIR grant, or otherwise fund early start-up activities.

Another often effective strategy is a strategic relationship with a larger company. Although sometimes a trap for the unwary, a carefully negotiated partnership agreement that grants narrow rights to IP can be an extremely efficient way to achieve progress.

A frequently ignored strategy that is the most non-dilutive is revenue. While the side of the road to glory is strewn with companies that thought they'd make some easy money selling diagnostics to finance their drug development, there are a fair number of companies that have found profitable near-term products that have been used to finance much more profitable later developments.

Regardless of funding strategy, the critical factor is to demonstrate progress toward the company's product development milestones. In the biotech industry, early stage progress is usually defined in terms of steps toward regulatory approval. A milestone-driven development plan plots the company's course to an IND (Investigational New Drug) and helps navigate through early clinical trial phases. Therefore, efficient use of funds to hit expected milestones along the regulatory pathway greatly enhances the probability of success. The extent to which a company can meet its early milestones with little or no venture capital will also greatly enhance the probability of the founders and early investors making a good return on their investment.

Spotting Opportunity

Although there are a large number of venture funds on the two coasts willing and able to make seed and very early stage investments, we are among a relatively few partnerships in the Midwest who typically make such early bets.

As such, we are willing to work with university tech transfer offices and corporate spinouts to actually take technologies that are pre-business plan and pre-management team and construct a deal around them. Although widely regarded as extremely risky, such investments are amenable to the same disciplined investment process that governs later stage deals. Statistically, seed and early stage investments have always been the most profitable venture sector.

Deals at such an early stage are obviously extremely time-consuming and most investors can only put together one or two each year. In such a labor-intensive hands-on environment, the leverage of a network of managers, consultants, and specialists is crucial. This is more important and more difficult in underserved markets, and especially so in biotech, but in a region such as the Midwest with a vast network of academic institutions, health care companies, and experienced lawyers and accountants, it is not difficult to find the necessary skill sets.

Like most venture capitalists, we do not assume that we are smart enough to be the only guys in the game. If we are unable to find an experienced co-investor, then it is highly probable that we should not be there either. Although it has traditionally been difficult to syndicate bioventure deals in the Midwest, in recent years a group called the Mid-America Healthcare Investors Network has been formed. MHIN is an association of more than forty-eight venture capital firms from fourteen states, with more than $2 billion under management. MHIN focuses on life science investment opportunities in the mid-America region. These life science opportunities include companies involved in biotechnology, medical devices, bioinformatics, healthcare information technology, and healthcare services. The group gathers quarterly to facilitate co-investment opportunities and is a valuable resource for the venture business in this less bio-investor-rich area of the country.

Growth Potential

Like most things in life, biotech investing is a search for balance. Blockbuster drugs that big pharma increasingly rely on for sustenance tend to be more difficult and expensive to get through the regulatory pipeline. Orphan and fast-track products, while "faster" to market, are definitely smaller.

Paradoxically, while the cost of taking a new pharma product through the regulatory system has continued to increase alarmingly, the failure rate, especially at later phases in the process, has not declined. As a consequence, the cost of taking a product to market has exceeded the reach of almost all but the largest bio and pharma companies. An increasingly sophisticated stock market observing these trends is decreasingly willing to finance such high-risk endeavors through the traditional IPO route.

So early stage bio investors have begun to see their role as more of an intermediary, bridging the gap from academia where most real innovation takes place to much larger bio/pharma companies that are financially and experientially equipped to take products all the way to market. Some good news is that more and more academic institutions see themselves as having a role in early product development and big bio/pharma is increasingly vocal about the need for them to move down the regulatory pipeline for acquisitions.

Evaluating Return Potential

Evaluating the potential return on a biotech investment can be a complicated task. We frequently get investment bankers involved before we make the investment. Given their specialized knowledge of the marketplace, we ask them to look at the deal and provide some idea about return potential based on the status of the market. Essentially, you want to determine how long it will take before the large companies take notice and move in. Because pharma's pipeline is perceived to be empty these days, there is an increased willingness for acquiring companies to look earlier and earlier.

As more and more blockbuster drugs reach the end of their patent life, big pharma is accelerating its process of growth through co-development partnerships or acquisition of biotech companies and their technologies. The nature of such M&A activity is increasingly moving earlier in the drug development timeline. Just a few years ago, biotech could not have expected a deal with big pharma before Phase III. Today, exit opportunities may present as early as the preclinical phase, although more likely during Phase I or II clinical trials. In fact, it used to be the case that pharma would never look at a technology until it was in Phase III, and now little is available in Phase III that is attractive.

Risk Evaluation and Management

As we evaluate opportunities, my partner asks about categories of risk: What is the management risk? What is the financing risk? What is the technology risk? What is the marketing risk? So you might mitigate financing risk by entering into the deal with a syndicate of other investors, who, if not individually, at least collectively have deep pockets. The management risk, of course, is minimized by looking for management teams with strong track records.

Frequently, the technology risk is the most difficult to minimize. It is extremely challenging to make a judgment about a technology's soundness or potential. It often amounts to a bet, one that you can make assessments about odds, but which you can never say for certain is a sure thing. Sometimes it looks like the technology is a slam dunk, but there remains a regulatory risk. How far can you get it down the pipeline, and can you do that efficiently with a relatively small amount of dollars and still get a decent return?

At every stage of the product development timeline, you are confronted with new regulatory risk—particularly with regard to new drugs or biologics. After completing preclinical testing and filing the IND application, you face the uncertainty of translating safety from animals to humans during Phase I. Entering Phase II you encounter the risk of maintaining safety as well as efficacy in targeted patient populations. Even if the investigational technology passes to Phase III, the question of whether the benefits exceed that of the current standard of care remains.

Furthermore, at every stage of the regulatory process, the company runs the risk of mismanaging trial protocols or not effectively satisfying regulatory requirements. In these cases, entire studies or portions of studies may require repeating, resulting in the need for significant amounts of additional funding and time. Like other regulatory bodies, the FDA is responsible for adhering to certain guidelines; however, the approval decision is essentially a judgment call. Again, biotech investors often look for Orphan Drug or Fast Track indications, which facilitate closer and more frequent interaction with FDA officials throughout the regulatory timeline, minimizing inefficient use of time and resources.

Especially in the case of the ever-evolving biotech industry, ambiguity contributes to regulatory risk. New guidelines addressing the concerns of new technologies may be issued at any time. For example, guidelines affecting generic drugs, as they apply to small molecules, have been in place for years. With the entry of follow-on biologics (i.e., "generic" biologics), companies as well as the FDA are realizing that the same rules cannot apply. Therefore, these companies face new regulatory risk as guidelines are not yet fully or clearly defined.

The message remains—the sooner that big pharma becomes interested, the less technology and regulatory risk the company must manage independently.

Evaluating Technology

Like most venture firms, we have a large network of contacts. We do not have an in-house department of experts, and we have yet to be persuaded that you can have enough of the right kind of experts in an increasingly specialized world. There are no longer any truly encyclopedic minds, so you *need* to have access to a diverse external network.

Patent lawyers can tell you not only about the strength of patents—their specialized expertise and familiarity within a particular field may help define an appropriate market entry point, for example, or identify new competition. Maintaining a network of other venture capitalists, industry executives, entrepreneurs, consultants, scientists, professors, and corporate and patent attorneys, among others, offers a means of accessing information and opinions from the right person. Moreover, if you happen to find a hole in your network, chances are that it can be filled by a simple referral from someone in your network. This extended network is essential in thoroughly and accurately evaluating technology.

Since we primarily restrict ourselves to university and corporate spinouts at an early stage, there is an intrinsic amount of technological risk reduction. We deal with first-rate research universities or institutions that have typically done the early stage patent work and invested enough to give you some confidence that there is "something there." Remember that universities do not patent all invention disclosures. Therefore, technologies we see have already made it through the first pass, having been selected by the university as a worthy investment.

Additionally, there is some assurance knowing that you are dealing with peer-reviewed technology invented by well-regarded scientists. Many technologies have already made it through the rigorous evaluation of a leading peer-reviewed journal. These factors place university and industry technologies in a more comfortable position from the investor's perspective. Technological risk reduction is just one reason why we do not typically deal with the garage shop guys who may well have superior technology but simply involve greater risk assumption.

Assembling a Management Team

With pre-seed investing, the most imperative concern is finding a qualified CEO. You can and probably should outsource most other pieces of the team, but you really need a domain experienced and competent individual at the helm. Some are willing to invest in an academic CEO, with the expectation that he or she can get the product down the development pathway without expending as much money on outside resources. We have never found that to be a very effective model. Academics are good at what they do. They are creative and inventive, and, for the most part, they have freely chosen not to enter the industrial world, where the expectations are very different.

Drug development (as opposed to discovery) requires repeatable, robust, industrialized processes that will generate a marketable product. As we consider risk reduction, we have found that betting on a proven management team—one that has repeatedly taken products out of the lab and into the marketplace—has a higher probability of success.

Valuations

Our approach to valuation is based upon our typical investment strategy. Because we are dealing with very early stage companies, if we cannot invest very cheaply, it simply does not make sense to get involved in the company. We aim to invest a relatively small amount of money, say $1 million plus or minus in the first round, while maintaining investment control of the company. In sum, we will never give a valuation that is much more than $3 million, and more typically, the company will be valued in the $1 to $3 million range.

The art of valuation is the same no matter what stage of the process. As the venture capitalist, you will establish a series of milestones that you expect to be accomplished with the money invested within a certain amount of time. To the extent that the management team meets or exceeds those milestones within budget and time constraints, it has the opportunity to increase the valuation and thus be effective at raising money more cheaply in the next round.

If a company wants to increase its valuation from the start, the best thing to do is get as far along in the developmental process as possible without the involvement of venture capital. Even if venture capital is involved, the more a company can do with other kinds of non-dilutive financing, the better off it will be. For instance, if a company raises $1 million from VC and attracts three times that amount through the SBIR and STTR programs, and state and local grants, it will go much farther than it would have from the $1 million raised from VC. Moreover, the company's valuation will benefit accordingly.

Entrepreneurs are almost always misguided in terms of what they think the valuation of the company should be. To a large extent, this is a matter of simple naïveté—they just don't understand the venture capital game. We very often find that they really do not understand how we make money and what our financial structures and incentives are. There is an assumption that somehow magically we have money, and we are supposed to give it to a good cause that may make life better for sick people. Now, that may be true if you are Bill Gates, but venture capitalists like us with limited partners need to know the probability for a large and profitable investment return.

Our job as a venture capitalist is to keep valuations low and aim for the highest returns, and the sophisticated CEOs and entrepreneurs, who have been through this two or three times, understand that. That is why the really good entrepreneurs try to go as far as they can without venture capital, which is exactly what they should be doing.

Negotiating Deal Terms

At our firm, we are fairly conventional in terms of deal structure. We don't believe that draconian terms work to anyone's advantage; even if you succeed in imposing them on entrepreneurs and management teams, when it comes time to do an M&A exit, they will demand a renegotiation in return for supporting the transaction.

Although we typically take participating preferred stock, we do not put in preferences that would require a grand slam home run for the entrepreneurs to make any money. Keep in mind, however, that deals are stage related. Whereas you would probably include a weighted ratchet in a later round, you have to put in a full ratchet at a seed or Series A round.

We want to make sure that at the end of the day, when you count up all of the preferences, interest rates, and other provisions, we have limited the valuation to the $1 to $3 million range for a seed deal. If someone wants to argue the price side, we will narrow the difference with other terms and conditions as long as the valuation remains in range.

The tendency on the part of naïve entrepreneurs is to focus entirely on price, whereas what they should focus on is retaining as many rights as they can under the circumstances. We always advise them to obtain sound legal advice, as we want to avoid a situation where they are not as well represented as we are. Entrepreneurs should concentrate on keeping the negotiations and the deal itself as balanced as possible. Again, the conditions are going to be a function of what stage they are in.

In general, however, there is not a great deal of room for negotiation with entrepreneurs. Especially where there is not much competition, you are not going to waste your time negotiating at great length. It is essential that the venture capitalists as well as the entrepreneurs understand that *nobody* is going to get rich in the first round of investment. We make it clear that we want to be fair with them, and they need to be fair with us. We want both sides to focus on what each can efficiently accomplish with this investment and how far we can move the company ahead. If all goes well, everyone will get rich together in the end.

We have been in this business for a long time, and we have done these negotiations hundreds of times. Thus, our strategy is to be very open and up front at the outset. We very often will work with a group of entrepreneurs on an informal advisory basis for a while just so we can get to know each other. This allows us to socialize the way we like to invest. For example, we are very candid from the start that we focus at least as much on governance as on deal terms. Fundamentally, we want to be very sure that we have board control of the company, so that in the almost certain event that we need to replace the CEO, we have the power to do so.

As I think is true for most venture capitalists, we find that there are different types of CEOs for different stages. There are start-up CEOs, there are next stage beta/preclinical trial CEOs, and there are ramp-it-up CEOs. Depending on your focus, the guy/gal that got the ball rolling is quite unlikely to be the one that exits it. Thus, you want to be very sure that you can make that transaction as easy as possible.

Likewise, we make it very clear that if the entrepreneur is an academic, he or she will not be the CEO of a company we are going to invest in. If that is his or her dream, then we should not waste each other's time. On the other hand, if the entrepreneur is interested in partnering with somebody with serious industry expertise, we will help find an appropriate domain-qualified individual to fill the role as CEO. Of course, the caveat is that it must be someone we agree upon, or again, we are not going to invest.

Structuring a Shared Investment

One of the issues that venture capitalists love to argue about is whether and when to bring in strategic partner investors. In the Midwest, funds tend to be smaller and focused on earlier stage investments. We worry about sharp-elbowed "coastal investors"—deep pockets from the East and West coasts that will come in and crowd us out. Hence, to the extent that we can, we bring in strategic investors so there is no power imbalance between the deep-pocket investors. How successful you are at creating that balance is obviously going to make a difference in how the deal gets structured. Fundamentally, you are not very likely to change the basic structure of a venture deal. Deals are put together the way they are for a reason, and they essentially offer the right kind of protection to all investors.

Thomas L. Churchwell is managing partner of ARCH Development Partners LLC, an early stage Midwestern venture fund. Previously, he was president and CEO of ARCH Development Corp. and president and CEO of Calgene Fresh Inc. He held several senior management positions at the NutraSweet Co. and legal positions at G.D. Searle & Co., American Hospital Supply Corp., and The Coca-Cola Export Corp.

Mr. Churchwell holds a B.A. from DePauw University, a J.D. from Northwestern University School of Law, and an AMP from Harvard Business School.

Acknowledgment: *Tom extends his appreciation and gratitude to Caralynn Nowinski for her excellent assistance in writing this chapter.*

Caralynn Nowinski expects to graduate from the University of Illinois at Chicago in the Spring of 2007 having earned her joint M.D/M.B.A. degrees. Ms. Nowinski earned her B.S. Honors in Human Communication Sciences at Northwestern University. Caralynn's work experience includes Graduate Research Assistant at the University of Illinois at Chicago College of Medicine. Prior to that, she was a Senior Research Manager at Mount Sinai School of Medicine in New York, Research Coordinator at the Northwestern University Medical School, Chicago, and Research Assistant at Pediatric Neurology, S.C., Lake Forest, Illinois

A Random Walk Down Pearl Street

Kyle Lefkoff

General Partner

Boulder Ventures Ltd.

I have been investing in early-stage life sciences start-ups in Boulder, Colorado, since 1986. Over the past twenty years, the venture capital and biotech industries have grown up around each other, and Boulder has been one of the key places around which that growth is centered. Some of the lessons that we've learned are true for all startups; some are true specifically for companies that try and address large unmet medical needs; and some are true only for companies that grow up in Boulder.

Our goal at Boulder Ventures is to make really original mistakes, and investing in cutting-edge biotech companies gives us lots of opportunities to do that. We try to avoid unoriginal mistakes by investing in experienced serial entrepreneurs; by forming investment syndicates of experienced investors like ourselves; by focusing on business models that we understand and that work; and by trying to back companies that help cure human diseases, which is a big problem.

Backing seasoned management is a hallmark of great investors, and the entrepreneurs who consistently outperform their peers in starting new companies are highly prized among venture capitalists. In the life sciences, these characters tend to have distinguished backgrounds in chemistry or biology, an unerring nose for the commercial potential of new science, and a deep understanding of the economics of the pharmaceutical industry. Novice entrepreneurs are frequently astonished at the speed with which such individuals can assemble a professional management team and a well-heeled syndicate of investors, but there is no secret in the VC world as to the value and importance of these serial entrepreneurs in creating lasting value among our portfolio companies.

In addition to experienced management, a proven syndicate of investors represents a crucial strength to start-ups generally, and to biotech companies in particular. Pharmaceutical development is a capital-intensive business and an experienced group of established venture funds can provide more money, credibility, and contacts that any one firm alone. A proven syndicate is also more likely to make original mistakes as a group, since they've been down this road before and seen the obvious perils. Finally, it's nice to hang out with your friends, so a proven syndicate can enjoy themselves and focus on the important issues that face every start-up, instead of trying to impress each other at the expense of their investment.

Successful serial biotech entrepreneurs focus their funding efforts on proven syndicates; all know that the combination of the two is an unbeatable advantage.

While the importance of serial entrepreneurs and proven syndicates has been an immutable feature of successful biotech companies for two decades, the drug discovery paradigm has evolved over time with the growth in the biotech industry and the development of its enabling technologies. For much of the post-war era, pharmaceutical companies devoted themselves to the identification and synthesis of orally active small molecules that addressed known biochemical targets in large populations. Vaccines, antibiotics, hormones, antipsychotics, and statins became the basis of a half dozen multibillion-dollar global pharmaceutical franchises.

Because these targets were well understood, discovery efforts tended toward industrial scale screening exercises, slogging through millions of chemical variants to find the best molecule. And because small chemical differences have profound effects on drug efficacy, successful discovery organizations focused on incremental improvements and were highly secretive. With the exception of Syntex, a pioneer in hormone synthesis, all of the most successful global drug discovery firms from the 1950s through the 1970s grew by acquisition and internal product development, rather than as start-ups financed by venture capital.

But with the advent of expressed human proteins in the 1980s, the rules changed. There emerged a new model of drug discovery focused on external academic collaborations, large molecules, and novel platforms funded by venture capitalists. We're twenty years into this model now, and whether biotech investors achieve liquidity through an IPO or through a sale to the large pharma companies, it is clear that venture capitalists have created a sustainable alternative path to building large pharmaceutical companies with this collaborative approach.

The allure of building drug companies for professional investors is powerful today, not just because the economic rewards for success are great. Equally strong returns are available to careful investors in early stage software, real estate, or energy companies. Instead, life science investors have the opportunity to realize venture returns and to cure human disease, a compelling proposition to those of us who want to improve the human condition rather than just our own.

Kyle Lefkoff is a founder and general partner of Boulder Ventures Ltd. Boulder Ventures manages four venture capital funds with more than $250 million in committed capital. In his twenty-two year career as a venture capitalist, Mr. Lefkoff has invested in fifty-four Colorado companies, resulting in twelve IPOs and fourteen trade sales.

Mr. Lefkoff's representative life science investments include Hauser Chemical Research (NASDAQ: HAUS), NeXstar Pharmaceuticals (sold to Gilead Sciences), Array BioPharma (NASDAQ: ARRY), Dharmacon (Sold to Fisher Scientific), and Genomica (sold to Exelixis). Representative information technology investments include McData Corp. (sold to EMC), Atlantes (sold to Metromail), Compatible Systems Corp. (sold to Cisco Systems), and Wall Street on Demand (sold to Goldman Sachs). Representative natural foods retailing investments include Horizon Organic Dairy (sold to Dean Foods), Alfalfa's Markets (sold to Wild Oats), and Fresh & Wild PLC (sold to Whole Foods).

Mr. Lefkoff is a director of Centerstone Technologies, Metabolite Laboratories, and Trust Company of America. He is the chairman of the board of Array BioPharma, ARCA Discovery, and BaroFold Inc. He is the past chairman of the Venture Capital Association of Colorado and is an adjunct professor of finance at the University of Colorado School of Business.

Mr. Lefkoff received his B.A. in economics from Vassar College (summa cum laude) in 1981. In 1980, Mr. Lefkoff completed a fellowship in economic history at the London School of Economics. In 1985, he completed his M.B.A. in finance at the University of Chicago.

Mr. Lefkoff is an experienced alpinist and was a member of the 1986 American Expedition to the North Face of K2. He is a life member and serves on the board of directors of the American Mountain Guides Association.

Dedication: *This chapter is dedicated to Professor James H. Lorie (1922-2005) University of Chicago Graduate School of Business, who was my teacher, mentor, and friend.*

Negotiating Points in Venture Capital

Laurence A. Lasky, Ph.D.
General Partner
Latterell Venture Partners

Background and Basics

I came to the biotechnology industry from the science side, and thus it is not surprising that I have a scientist's outlook on the industry. For me, the goal of biotech investing is the development of drugs and products that are clinically important. In order to increase the probability that this will occur, I look for the best science and scientists. Such an approach might make for a more challenging road, particularly at the outset, but it will usually end in success that is both gratifying and lucrative.

I have learned to look for science that is backed by a significant amount of literature in biochemistry, genetics, and molecular biology, along with a great team or a great leader to push it along. Of course, biology is a complex field, and sometimes, even with solid ideas and strong personnel, things do not work out the way one would hope. However, the most you can do as a venture capitalist is maximize your potential for success, and that starts with good science and motivated people.

I take a close look at who the founders are and where they have come from, and only if that group passes muster will I then take a long hard look at their idea. Very often, people are extremely strong academically, but drug development is very different from academics. In academics, one might start out researching one thing and along the way discover something else that is interesting and follow that new lead. That kind of opportunism is usually not a part of drug development, which is a much more focused process with very specific goals. One really must concentrate on the objective, which, of course, is to get the drug into people and see if it works. It is important to try to evaluate whether the developers have talent that extends beyond academics into application.

Part of the evaluation process is understanding the field. If I am not very familiar with the area, I will read a tremendous amount of background literature including reviews, academic papers, etc.—as well as speak with experts and do whatever I can to acquire the necessary level of understanding. This is a crucial and important step in assessing the likelihood of success. I think the one advantage in our group is that we have people who have directly worked in the industry with their own hands. For instance, I have about 135 publications and thirty-five issued and pending patents, which gives me an insider's perspective on the entire process.

Managing Personnel

The most challenging part of the venture capital game for me is in the interpersonal aspects: i.e., dealing with all of the different personalities involved in a way that is effective while not being overbearing. For instance, in academic science the way it works is that I will essentially be in charge of a laboratory; there is certainly a give and take, but I am the boss. I pick and choose who I want to work on the project, and it is my responsibility to direct them as best I can. Striking the right balance between helpful direction and suffocating control is a subtle and complex problem. As the individual in charge, I must keep the project members focused, while allowing them room to function effectively. The management of people in the venture business involves much more subtlety and sensitivity to the individual ego and insecurities. In addition, because I only usually hear about results at board meetings, I need to trust the management to perform in the most rigorous manner.

Intellectual Property

Other less difficult problems are rounding up the intellectual property rights, which usually is a relatively simple process. However, there have been instances where a company has gotten all the way to the final stages before I discovered that there was a significant problem with IP, in which case things can fall apart very rapidly. A small start-up has no protection against the Genentechs and Mercks of this world beyond their intellectual property. Because it requires the services of a lawyer, IP analysis can be quite expensive. We usually hold it off until the end of the process, because it can cost tens of thousands of dollars. Of course, if the funding of the company occurs, than they will pay us back for those costs.

Growth Potential

In selecting a project, we look for developments that will address unmet medical needs. With drugs, it is useful to avoid "dirty drugs" that have a high degree of off-target activity. We like to look for drugs that have a cleaner and more specific impact on the target itself.

Other things to look for are novel targets with a huge amount of biology behind them, supported by a significant amount of literature on potential targets and indications. I have a preference for oncology because the side effect issues are much less significant. People will put up with a lot more side effects when dealing with a lethal disease. The success barrier is also lower for cancer. If you give a cancer patient a few months of progression-free survival, that is a home run. There are many new targets in the oncology area (cancer stem cells, for instance), and there is a lot of opportunity for new products and new companies.

Risk Assessment and Management

Drug development is very high risk for two simple reasons. First, there is the risk associated with toxicity. Second, there is the question of efficacy. Whether you can make a drug for a particular target that will ultimately hit that target is not actually the main issue. A more important issue is whether, when you first put that drug into animals, it shows significant toxicity. If that is the case, you have to think very hard before putting it into human oncology applications. With regard to the efficacy question, most of these targets are first in class, meaning they have never been done before. With most of these targets, there is really nothing out there other than genetics and biochemistry, so you need to understand the science and read the recent literature on these targets. In this manner, one can enhance one's chances of success, although the proof is still in large-scale clinical trials.

A less dangerous but still significant risk is the actual time to development. Everything takes longer and costs more than you think, and the question ultimately is whether the syndicate will have the financial stamina to put up with the increases in time and cost? The way that I try to get around that is to make sure that the science and the team are very exciting, so that even if things do not go perfectly and the syndicate has to wait longer and come up with more money, they will be willing to do so because of the sheer potential of the science.

The Team

A successful biotech deal involves contributions from a number of different parties. At the top of the list, obviously, the company founders are incredibly important to the deal. Sometimes they are serial entrepreneurs who have done this before and sometimes they are new to it, but the most important thing is to get involved with people who are extremely smart and come from very good universities who will understand that drug development is a long, difficult, and costly task. Being academics, they can have rather prodigious egos and can occasionally be difficult to put up with, but, in general, if you get involved with people who are first-rate academically, it is *always* a huge plus. I have started successful companies out of Yale, Cal Tech, Princeton, and the University of Michigan Medical School, where the people involved are academically excellent and that has been an important aspect to the success of the companies.

Second in the hierarchy of importance is the scientific leader of the company, the CSO. The CSO has to be a very experienced biotech person, and he or she must have a realistic mind-set as far as the difficulty of accomplishing preclinical and clinical goals. The CSO has to be extremely honest when things go poorly. We want to and need to hear the bad news along with the good news. He or she also has to really understand the science. Typically, they should be somebody who could have been a professor at a major university. The CEO obviously is another hugely important member of the team. An effective CEO must be very experienced at fundraising, has to have a tremendous amount of charisma and leadership ability, and has to be someone whom we can believe in to get things done and push things. Of course, it helps to have a CEO who can think very strategically and who is prudent about spending (not flying around in private jets, for instance). The CEO needs to be focused on what we really care about, which is getting the drug in the clinic.

Further down the line, but certainly of great important as the project moves forward, are the drug development people. Once you have a target and a lead and are ready to go to the clinic, then a person who has done drug development in the past is extremely valuable. Such a person understands how to do preclinical tasks, understands how to file INDs without raising any red flags at the FDA, etc. Finally, but not to be overlooked, is the legal team. Lawyers, particularly IP lawyers, can be either in-house employees of the company or outside consultants. In either case, with biotech development it is crucial to have lawyers who are very sophisticated with patents, who understand that getting broad claims is a necessity, who know how to deal with the patent office and other lawyers, and who understand science.

If you have strong people in all of these areas, they tend to attract even more good people to the company. This is how Genentech was initially successful. They started out with the first good group of people and more talent naturally gravitated toward the company.

I believe that my personal role is much more scientific than the average venture capitalist. I am not overly interested in the financial aspects of a project; there are others in our group who are really good at spreadsheets and financial strategy. I am much more interested in how the science is progressing, what experiments are being done, how I can help them think about experiments, forwarding along papers or other literature that I come across that may be relevant to the project, etc. My input is almost 80 percent science. It is what I enjoy and it allows me to feel like I am still working in the field. Of course, I understand that we are supposed to be making money for our LPs, but my number one motivation is to make drugs that will treat diseases, and if we do that, then the money will come.

Negotiating Points and Strategies

I am usually quite involved in negotiations. It is not where my interests most keenly lie, but negotiation is an important part of the process. For the most part, the two sides are usually not that far apart, and so the negotiations tend not to be overly difficult. The term sheet negotiations, in fact, are more like an education process, because many of these founders really do not have a firm grasp of what the sheets contain. Once they understand the terms, they are typically fine with them. In general, they tend to appreciate the fact that you are spending time with them and that you want to help them out. They understand that they have an opportunity to make some money, and that you are there to help them succeed with their new technology. Of course, in my case, the fact that I have a background in science certainly helps. The founders are often familiar with me beforehand,

and I think they tend to trust me more because I am "one of them." They tend to look at me as a fellow scientist trying to help them develop their technology. When I negotiate, I am not trying to manipulate people.

Pre-Money Valuation

The most significant negotiating point is usually the "pre-money," or what the value of the company is before the funding. Our rule of thumb, which actually seems to work, is that a preclinical company coming out of a good lab with compelling data with an indication in a market that could be quite large is worth between $3 and $5 million (a number that includes the founders' shares as well as the option pool).

At our partnership, we do almost strictly A rounds, meaning that we are involved with the project from the beginning and that the valuations will be rather modest. It is a significant amount of work to do an A round, because in essence you have to put everything together yourself. However, the flip side is that you can get these companies for very little money, as the pre-money valuation will almost always be rather low (again, for us, usually less than $5 million). A B round is actually much more complicated, because there is lots of baggage, everything is in place, the company has had a lot of issues and solved some and not others So the valuation of B rounds can vary dramatically, and the negotiations are usually a bit trickier. On occasion one gets series B term sheets, for instance, where the first round investors have included protections like full ratchets and the like. These are a little more painful to deal with, because they protected themselves from down rounds, which is what the Series Bs often are.

With an A round, it is much simpler as far as the risks and rewards. That $5 million or less valuation includes a big option pool. In general, we want an option pool that is 20 percent of the total shares, so that we have enough shares to give to the CEO, the CMO, the CSO, plus all of the employees. It is important for entrepreneurs to not have an unrealistic notion of what the company is worth. The A round is unbelievably risky, with the scientists usually having some animal data and significant supporting literature. A huge number of things can still go wrong at that point. This is step one in the thousand steps of drug development, so an entrepreneur should not expect an A round valuation in the tens of millions of dollars.

Sometimes the founders insist on a board seat, which usually is something that we give them, along with the proviso that at some point we will have to change to a more professional board member. That can be a sensitive issue, but usually one can get the founder or founders to understand that a highly professional board is what is best for the company.

Milestones

Another negotiating point will be the amount of capital that a company needs to achieve their milestones. This is obviously extremely difficult to predict due to the difficulty involved in turning the academic or the prototypical into the applicable. Of course, we always insist that the founders and the CEO come up with a budget, but such a budget can be far from a dependable indicator (in fact, they can pretty much get the budget to say what they want it to say). It is difficult to predict with any certainty how much money will get them to certain milestones. Usually, it makes sense to err on the side of giving the founders a little bit too much money to be sure that they get to where they want to be. The board needs to watch very carefully as the company develops to be sure that there is enough money and that it is being spent efficiently. All funding deals now are "tranched," meaning that only a certain percentage of the money is given up front, with the rest being disbursed after the company has achieved a certain milestone (usually filing an IND and treating the first patient). To get the rest of the funds, the company really needs to accomplish something significant. This helps them to focus on the objectives and it helps on our IRR because we are only giving them a fraction of the money up front.

Negotiating Success

There are a few simple parameters that would lead me to label a negotiation "successful." First and foremost, the pre-money should be low enough so that the amount of capital we are putting into the company gives us a significant ownership percentage (in general, we like to be 10 to 30 percent, depending on the company). Also, it is important to have a reasonable budget and to have the funding scheduled in such a way that the company must really reach a significant milestone for not a huge amount of money. The less money needed to reach that milestone, the more of an uptick the company will see in its valuation in the next round. That, in the end, is what one wants: for the company to take a decent step up, so that everyone will be rewarded for the hard work they have put in.

Advice for Entrepreneurs

It is important for entrepreneurs to focus more on the team of venture people that they are dealing with than on any initial dollar valuation amounts. Go into business with a venture group that knows what it is doing from both the science and financial viewpoints. The single most important factor is the quality of the venture group itself. You want to work with people who are smart, strategic, know the drug development business, and can find people that can help you when things get difficult.

Laurence Lasky is a founding general partner with Latterell Venture Partners. He is an expert in biotechnology, a field that he has worked in commercially for more than twenty-three years in addition to his early pioneering scientific work in the field beginning in the mid-1970s.

Prior to joining LVP, Mr. Lasky was a leading scientist at Genentech for twenty years where he worked in various disciplines including vaccinology, immunology, stem cell biology, cellular signaling mechanisms, and monoclonal antibody therapy of tumors. Two of the vaccines that he worked on, for herpes simplex virus and HIV infection, are currently in phase III clinical trials. He is a co-author of more than 131 publications and a co-inventor of twenty-seven issued patents, several of which generate significant licensing royalties. He was a standing member of the Research Review Committee which directed research across all product development areas at Genentech.

Mr. Lasky has also served on numerous NCI and NIH scientific review panels and has reviewed many papers for a number of prestigious journals.

Mr. Lasky holds B.A. and Ph.D. degrees in biology from UCLA and completed his post-doctoral studies at The California Institute of Technology.

A Deal is Just a Deal

Lawrence R. Goldfarb

Chief Executive Officer and Portfolio Manager

LRG Capital Group LLC

A Vision of Venture Capital in Biotechnology

The venture capital industry has existed since the beginning of time, in one form or another. In essence, an entrepreneur identifies an issue and attempts to address it—to do so requires an infusion of capital. There are two primary sandboxes in which venture capitalists play: technology and life sciences. Technology encompasses a broad array of industries, including enterprise software, hardware, wireless connectivity, and digital media. Life sciences consist of two primary areas: biotechnology and medical technology. Biotechnology is primarily concerned with the development of therapeutic compounds (drugs), whereas medical technology is comprised of a wide array of medical devices such as pacemakers, bionics, and diagnostics tools.

The primary difference in the technology and life sciences sectors, besides their obvious industry focus, is the regulatory approval required for products of life sciences companies. The Food & Drug Administration (FDA) imposes strict requirements on both the biotechnology and medical technology sectors. The exhaustive research and development required to meet these requirements causes life sciences companies' business models to move at a much slower pace than technology companies. Of the two life sciences subsectors, biotechnology proposes the most stringent requirements; hence, the balance of this chapter will focus on biotechnology.

Biotechnology, by definition, involves the application of certain biological substances to perform specific processes. Typically, the venture capital industry uses the term "biotechnology" as a narrow categorization to refer to entities involved with the development of some form of medication.

As life expectancies continue to increase and health care infrastructures persistently improve in developed nations, the global concern for medical care and the resulting interest in biotechnology life sciences continue to experience rapid growth. In addition, what used to be recognized as third world economies can now afford sophisticated life science technology, further stimulating the growth of the industry and resulting influx of capital. The magnitude of deals is increasing and competition for a part of the action is becoming more intense. The move to national health systems has resulted in a dramatic change in the way our health professionals are reimbursed by insurance companies, which has also contributed to the growth of life sciences investing.

For a company to be successful and generate interest from the investment community, it must uniquely address an issue that has large market potential. An unfortunate consequence of this prerequisite is the tendency for companies that are researching "small market" diseases to be ignored. This may seem unfair to some, but investors are in the business of generating returns, so it follows that the companies to receive funding are those that address widespread problems with inroads to substantial markets. The products that receive funding are those that concentrate on these widespread issues that affect a significant portion of the population.

A major hurdle associated with investing in the biotechnology industry is gaining an understanding of the science. How groundbreaking or revolutionary is a prospective investment's science? Investors require a fundamental understanding of the technology: Does it work, how expensive is it to manufacture, how is it distributed?

The next challenge is associated with the regulatory approval process. The products produced by the biotechnology field are heavily regulated by the FDA. Successful completion of the different phases of clinical trials associated with the approval of a new product is completely reliant on a board of FDA experts. Consider a hypothetical miracle drug that has the ability to cure cancer. Administration of the drug costs $250,000 per person per year and there is no way to reduce the costs. In this case, the cost poses a significant issue: Insurance companies will not pay for administration of the drug, regardless of its benefits, due to its costs. Therefore, an investigation into cost-reduction technologies must be undertaken. In addition, the likelihood and timeline associated with regulatory approval must be considered.

This regulatory process presents a key differentiator between biotechnology and the technology sectors of Silicon Valley: Apart from a minimal set of regulations associated with electromagnetic emissions and consumer safety, the conventional technology industry is almost completely unregulated. In the traditional technology space, regulators are not responsible for determining whether a product can be sold—this decision is made by the market itself. Suppose our hypothetical miracle drug cures cancer with an astonishing success rate and no side effects. Even so, the FDA has the ability to make obscure decisions affecting the drug's approval, citing the need for further study or even a repeat of a certain phase of trials. The FDA's power over the commercial viability of a biotechnology product is a huge issue that must be understood.

To help understand the science, it is prudent to retain consultants with specific knowledge of the technology under scrutiny. These consultants bring with them a wealth of knowledge resulting from working in their field for the majority of their careers. They are equipped to obtain an in-depth understanding of the product and issues associated with its commercialization. Due to the unique nature of each product, there is no one person with the ability to understand all the innovations coming across a venture capitalist's desk. And besides the science, the size of the market, distribution specifics, and competing drugs must also be assessed. One of the main advantages of using consultants for industry-specific due diligence is the fact that consultants do not have a vested interest in the completion of the deal. A consultant will understand their particular niche market and receive a fee regardless of the outcome of the deal, thus they are completely unbiased. Conversely, an in-house technology expert may be biased by his or her opinion of the deal or the opportunity to have his or her name attached to a groundbreaking innovation. This bias is avoided via the use of external consultants.

Regulatory experts with a proven track record of dealing with the FDA also play an important role in the deal process. We typically prefer to use experts ranging from consultants to lobbyists and lawyers who deal with the FDA's approval process on a regular basis. Incumbent conglomerates such as Johnson & Johnson typically have these people in-house. Small companies involved in funding deals, however, usually have limited experience with the FDA approval process. Thus, these regulatory consultants play a key role in the assessment of a deal. As with technology consultants, not only do these regulatory experts provide key insight into possible roadblocks associated with a product, but they also provide an unbiased opinion of the company's likelihood of success from a regulatory standpoint.

Identifying Opportunity

Understanding the science and market potential of biotechnology innovations makes it possible to assess venture capital investments as they appear. As mentioned previously, the initial assessment of a prospective investment is based solely on its market potential. The number of people afflicted with the addressed condition plays a key role in the investment process. Fortunately, data on medical conditions is readily available, as is an analysis of the issues with existing treatments, recovery probabilities, and life expectancies of those infected. All this information is used to provide an assessment of market potential. A venture capitalist has a fiduciary responsibility to produce returns for investors, therefore only investments with a significant market potential are typically pursued. A regrettable consequence of this line of reasoning is the obscure diseases with small market potential that fall by the wayside. These afflictions must rely on charitable organizations and grant-funded research for treatment breakthroughs, not venture capitalists whose primary concern is investment return.

Consider Lipitor, for example. Similar to the way in which taking a baby aspirin every day can forestall cardiovascular problems with almost zero side effects, Lipitor addresses a colossal market because it effectively lowers cholesterol with no downside (other than potential liver problems that can be addressed with simple tests during the first twelve months). Lipitor is an excellent example of a drug that has moved from the treatment space to one that is also used for preventative purposes. In addition, the age of patients to which the drug is prescribed continues to lower, as the drug is increasingly being administered as a preventative measure for younger patients. Lipitor presents an innovation that embodies the market potential sought in all prospective investments.

Following the identification of an investment addressing a market with acceptable potential, we examine a number of factors indicative of the company's outlook. A key component is the management team. Of particular importance is the management team's understanding of the steps required for successful transition from a research and development outfit to a viable operating business. Consider a late stage company management team with proven business know-how. The decision-making process associated with this type of investment is much more straightforward than one involving an early stage research and development shop. I have personally invested in billion dollar biotechnology companies with completely business-oriented staff—staff that have demonstrated the ability to transition a company from research and development to profitability. They understood the steps required to move from just spending money to realizing profit. With some management teams it is evident at an early stage that they're purely research and development outfits—they don't have the knowledge or experience to make the commercial transition and it's obvious the person in charge will never give up control. It's a no-brainer to walk away from a management team with these characteristics.

I've learned from personal experience that even though a company may have developed a revolutionary product with huge market potential, a CEO who is not an executive but an entrepreneur can be a real detriment to the company. An entrepreneurial CEO who has no knowledge of the steps required to move from technology shop to profitable business can destroy an opportunity that otherwise had a great deal of potential. Management needs to be willing to employ professional managers, because research and business are two completely different realms. It's a tough transition to take a company from research to commercialization and profitability, and much of the risk to an investor is associated with this focus shift.

At this point in the process, it's important to step back and consider the valuation model. We're essentially constructing a model that incorporates assumptions built upon assumptions, a process that is fraught with a number of pitfalls and shortcomings. It's evident that alteration of any one of the assumptions could easily change the valuation result, and in most cases the realized outcome will be different than that predicted by the model. Therefore, any modeling results are typically valued with a very high discount rate to incorporate not only investment risk but also the risks inherent to the model itself. It goes without saying that even though the model may not provide a highly accurate valuation result, it does provide the foundation for an iterative process that can in turn be used to generate cash flow projections. The modeling approach enables an investor to adjust assumptions and visualize the changes' affects on valuation. We can adjust assumptions concerning cash flow and sales, costs of revenue, and pre-cash flow, and observe the sensitivity of the company valuation to each of the inputs. The modeling approach and a sensitivity analysis enable assessment of the investment's risk as well as the potential opportunity; they play an important role in the decision-making process.

On a more personal note, investing in biotechnology ventures is all about the science, the market for the technology, the cost, and the management team. When it comes to meeting with individuals seeking capital, most of those I meet are scientists or Ph.D.s, researchers who are extremely passionate about their work. Their work is their baby, it's their life. Many of the researchers I meet don't see the forest for the trees, instead they see the bark and think about what bugs could be crawling underneath. They don't have a grasp of the big picture when it comes to building a profitable business. Researchers who have been living their research for a decade or more are often extremely accomplished in their niche field, but more often than not scientists don't make successful businesspeople.

In my experience, and this brings me back to my previous discussion about the importance of evaluating a company's management team, the entrepreneur is typically not the best CEO. This is why the best biotechnology companies I've seen often move the scientist/researcher into a scientific advisory or managerial role when the business is ready to transition from R&D to commercialization. This role is often known as the CSO, the chief science officer.

Structuring a Deal

There can be many players involved with the construction of a biotechnology life sciences deal. On the investor side, the parties involved range from individuals to venture capitalists to strategic investors. The latter usually takes the form of a company in a similar industry that is interested in the company's product innovations. Regardless of the investor's identity, there is ultimately a single person responsible for the investment—this individual is referred to as the lead investor. It's often the case that a syndicate of investors will be buying into a deal, and typically the syndicate will work together as they all have a common interest in the deal's success. The lead investor performs the bulk of negotiating and dictates deal terms, as the lead will usually be investing the most capital. In essence, the investor or syndicate is involved with prudently ensuring that the terms generate as great a return as possible while staying within the bounds of the law.

In addition, there is also a party responsible for memorializing the agreed upon terms of the deal. This responsibility typically falls to an attorney. It's very rare to see attorneys who specialize in biotechnology deals, as far as the legalities of the investment since there is nothing unique about the process. Effectively, the deal can be simplified to a purchase of some form of security in an entity that requires capital. An attorney doesn't require specialized knowledge of biotechnology deals to address the legalities of an investment. The attorney is responsible for protecting the rights of the investors, ensuring that the deal is adequately memorialized as well as acting to provide the investor with as many rights as possible. The attorney should also make certain that the investor gains a sufficient amount of control over the company. Typically, control does not become an issue until the company is in dire straits, at which point an investor may need to step in to take a lead role in the daily operations of the business. For example, a CEO's use of investor capital for the purchase of a personal luxury automobile or a company's failure to meet sales projections for four quarters in a row may dictate that an investor take hold of the reins. Remember, these are not usually very sophisticated businesses at the time of investment; therefore it's the lawyer's responsibility to obtain as many legal rights for the investors as possible.

Another important party is typically referred to as broker or intermediary. This role is often played by a company such as an investment bank. The broker is paid by the seller to contact investors and generate interest in the deal. The broker also negotiates for terms favorable to the seller, pitting investors against each other and bargaining with the syndicate. The broker is effectively a market maker, facilitating the deal without taking on risk themselves. In essence, the broker's job is to sell the investment as quickly as possible for a maximal price while devoting a minimal number of man-hours to the transaction. This modus operandi generates the optimal margins for the broker. If a broker can create an auction type process in which there are many bidders, then the broker effectively becomes a logistician because he or she is not directly involved in the selling process. The broker may create financial projections and presentations, the diligent investor will discard these in favor of meeting with the CEO and scientists firsthand.

Finally, it's obvious that the company, or seller, must also be involved in the structuring of the deal. Typically, the CEO is involved with the process as well as the board of directors, if one exists. While the CFO may also have a degree of involvement, the financial staff at early stage biotechnology companies is not usually well versed in the complexities of private equity investments. Thus, the head of finance will usually act more as a controller rather than CFO. The CEO, in my opinion, is the chief salesperson for the company. Therefore, not only does the CEO need to have a grasp of the company's science, but also a first-rate CEO possesses a fundamental understanding of the company's business. When the CEO understands only the science, then we are again faced with the rough transition from research to profitable business. A capable CEO will be able to explain why investing in their business is an attractive prospect, they should be able to convince you why your capital is better with their company rather than alternative investments.

In addition, if a board of directors exists, the board members can be very helpful during the deal structuring process. A first-rate board should provide a foundation for rationality during what can be a tumultuous deal structuring process. Each deal will appear differently, each proposal and term sheet will have pros and cons. The board can act as a key inside body with which the specifics of deals can be discussed, as well as being able to provide reasons for the pursuit or abandonment of certain proposals. A motivated board is a board with a financial interest in a deal's outcome. It is paramount to ensure that the board maintains a significant slice of equity and thus a vested interest in the company's success.

No two deals are going to be the same. Sometimes the only person I negotiate with is the broker, other times I have in-depth discussions with the CEO and board members prior to executing a deal. The following paragraphs outline my advice to the management team, to facilitate an understanding of the roles and motivations of all the players.

The only reason for a prospective investor's pursuit of a deal is return on their capital. Therefore, an accomplished management team is able to explain, as simply as if talking to a child, how the investor is going to realize a profit. An investor is not interested in hearing about a product that will save the world, investors aren't interested in charity. Investors want to know how much profit they're going to realize as a result of their investment in your company.

The management team should understand the individual needs of all prospective investors. A successful management team sits down with each investor and says, "Tell me about your firm, tell me how many deals you've done, what you've invested in, what's your investment approach, what are you looking for in a company, and what do you need from us?" Garnering an understanding of each investor's motivations up front can save a significant amount of time. Each investor's fit can be assessed within the first half hour of negotiations. I know from experience that management teams are often wary of undertaking this line of questioning at such an early stage, but the management team should be interviewing the potential investor, not vice versa. Even though management may feel such a forward approach could alarm an investor, experienced investors will be impressed by a management team that performs their own investor due diligence and is interested in maximizing the deal's efficiency for all parties involved. I'm always impressed by management that contacts other bankers and brokers with whom I've dealt with previously. It's encouraging to see a management team that is interested in whether an investor is a fit for their company.

Elements of the Deal

The most pertinent items negotiated in a biotechnology deal, in order of importance, are price, amount, and terms. Those are the three biggest items because in effect they determine how much control over the company an investor will realize.

Personally, I try to be realistic from the outset; I always fabricate a term sheet for the company that is realistic from the get-go. After initial presentation of a term sheet, the deal boils down to negotiations between the investor, the broker, the company, the board, and the rest of the investor syndicate. Negotiations with the latter will often see prospective investors pitted against each other in attempt to realize optimal terms for themselves. An experienced management team will have an arsenal of methods to maximize their company's valuation: sliding scales, valuation based on future funding rounds, valuation based on milestones, investments where capital is tranched and the release of funds is determined by the completion of certain key milestones. There really is no limit when it comes to the specifics of deal structuring and negotiations.

When it comes to negotiation strategy, I segregate investors into two categories: Investors who are only interested in maximizing their own profits, rather than having a vested interest in the company's success and the investor who has a sincere commitment to creating a positive outcome for both themselves and the company. In my experience, the former is a recipe for disaster. I typically structure a deal with realistic provisions, reasonable hurdle rates, and achievable milestones. This approach ensures that both the investor syndicate and company leave the table thinking, "I didn't get everything I wanted, but I like the deal," rather than one party leaving the table saying, "I overpaid" or "I got raked."

In general, my strategy does not typically modulate for each party involved in the deal-making process, however, variations do present themselves, depending on an individual's level of experience. For example, negotiations with an experienced broker may progress very smoothly and efficiently, while involvement with an inexperienced CEO may require a lot more effort as the deal will be a learning process for the chief executive.

Valuation

When it comes to valuing a biotechnology company, in essence I view the process as the construction of a very detailed, realistic, iterative model that provides sufficient understanding of the investment's framework and sensitivity to valuation assumptions. It is important to be able to assess the effects of your assumptions on the company's valuation according to your model. For example, if I have an assumption concerning the market potential of a company's product as an input to my valuation model, I may assign a 5 percent discount to that particular factor because as an experienced investor I feel I have an understanding of the investment's potential. When it comes to the success of the technology itself, I need to have an understanding of the risk associated with my technology consultant's input and discount the technology factor accordingly. As far as distribution and marketing of a company's product, I personally may not have the expertise to be able to address this factor. If Johnson & Johnson is going to distribute a company's product, then most likely the product will have an inroads into every doctor's office in the country and the risk associated with the distribution and marketing factor will be low. Every factor is effectively discounted at a different rate, each assumption is assigned a probability such that the model realistically values a company's cash flows while providing an insight into the model's sensitivity to assumptions.

Of course, the rate of return also factors into valuation. If the investment doesn't hit a certain required rate of return, I'll walk away from the opportunity. With a completed matrix of assumptions and a favorable valuation, if the rate of return is not sufficient then I will not invest. In my mind, the rate of return is a key factor involved with justifying an investment.

Entrepreneurs are often misguided in their interpretation of what their company's value should be. An entrepreneur will always consider their company to be worth more than an unbiased investor; I have never been part of a deal in my twenty-four years of investment experience where this wasn't the case. Entrepreneurs are one of the greatest by-products of America's open capital markets and innovative culture. Where else can you find so many individuals willing to mortgage their entire life for a dream? To secure business a broker will play on an entrepreneur's self-belief by leading the entrepreneur to believe their product is worth $100 when they initially thought it was worth $90. When an investor then values the entrepreneur's product at $10, it can often be interpreted as a personal affront, even though it's simply an unbiased evaluation of the product's potential. Brokers' tendency to play on an entrepreneur's self-belief in this manner only serves to complicate the deal-making process. This is true in virtually every industry, and biotechnology is no exception.

My strategy for countering the valuation difference between an entrepreneur and an investor is thorough preparation. You have to know the deal, be organized, have intimate knowledge of the company and their products. If an entrepreneur can demonstrate a first-rate management team, their company's valuation will be positively impacted. Realism is always an important part of any strategy. When entering into negotiations it's always important to have an experienced board of directors, a reputable broker, a diligent lawyer, and a realistic frame of mind. An entrepreneur must realize, and this may be contrary to the quantitative backgrounds of a great deal of scientists, that valuation is not an exact science. There are too many variables and assumptions for valuation to be an exact science. Companies can be valued within a general range, depending on industry and basic accounting ratios, much of which is based on publicly traded companies. There is a plethora of well-known rules of thumb used to value prospective investments. For example, given that there are twenty-four given public companies in a particular field trading at an average price-to-earnings ratio of ten, investors may use similar multiples to value a privately held investment in the same industry. Is this approach exact? Of course not, but this form of valuation provides a guideline, a single point of reference amongst a myriad of valuation methods.

As a diligent investor, if I've undertaken all the valuation activities discussed above, I will have a relatively accurate notion of a company's true value. I would have assessed a company's market potential, their science, the management team, and the probabilities associated with events affecting valuation. But even then, all of my effort only provides me with a guide—I won't be able to determine the accuracy of my modeling until I liquidate my position. I can only be sure of an accurate valuation upon realization of a profit.

Advice on Valuation

You must be realistic to pay a fair price. If you go into every deal aiming to invest at the absolute low, on most occasions you won't achieve your goal. Of course you don't want to overpay, but it's important to be able to walk away and say, "I paid a fair price."

A lot of investors will say to me, "I stole it," and then they discover a year down the road that all they stole is a problem. As I've discussed previously, investors who enter into a deal with only themselves in mind do not have a vested interest in the success of the company in addition to their own profit. It's important for an investor to enter into a deal as a partner to the company. I'm not promoting overvaluation, but it's important to have the company's interests at heart in addition to your own. As an investor, you don't want to lose face by having the biggest deal of the last decade in the headlines only later to discover that you grossly overpaid. You need to have an understanding of the market environment, factoring it into deliverance of your capital and the way in which the deal is structured.

A significant portion of company valuation boils down to common sense. Understanding the science, the market, the distribution of the science, and the cost of marketing and distribution are all established disciplines. But, aside from garnering an understanding of the specifics of the product and company, so much of a deal is just plain common sense and is applicable to life outside of finance. My point is, although it incorporates a great deal of specialized knowledge, there is no rocket science when it comes to structuring and executing a deal. Consider Warren Buffet, undoubtedly one of the most successful and well-recognized investors of all time. Warren Buffet invests in a myriad of different securities, but every investment is just another deal. The same goes for KKR, one of the world's oldest and most experienced private equity firms; KKR invests in a huge range of assets, but at the end of the day the bankers at KKR will tell you that a deal is just a deal.

Lawrence Goldfarb brings more than twenty years of investment banking, trading, and legal experience to LRG Capital Group. He is recognized as a leading expert on making direct investments in both private and public companies. His specialties are bridging the financial needs of growth companies by making direct investments and structuring transactions that enable them to grow and provide investor returns.

He began his career in 1984 as a tax attorney with the New York offices of Milbank Tweed Hadley & McCloy and thereafter Skadden Arps Meagher & Flom, where he provided tax analysis and consulting services in connection with a wide variety of complex transactions, including financings, mergers, and recapitalizations. In 1987, Mr. Goldfarb joined the San Francisco office of Credit Suisse First Boston Corp. as director of mergers and acquisitions. In 1992, Mr. Goldfarb joined GFC Capital Corp. (an affiliate of Loeb Partners) as president. Thereafter, Mr. Goldfarb joined Shoreline Pacific Investment Finance, a private placement agency firm, as executive vice president, before joining with others in 1998 to form BayStar Capital Management LLC, a vehicle in which he could fund growth-oriented, small capitalization, publicly traded companies.

Mr. Goldfarb regularly speaks at industry events and conferences, including the Institutional Investor's Hedge Fund Roundtable and the Alternative Investments Summit in addition to other private equity and high net worth focused conferences. Mr. Goldfarb has been a guest lecturer on arbitrage strategies at business schools, including Columbia and Berkeley, and has published numerous articles and books on investing and arbitrage strategies.

Mr. Goldfarb received his B.A. degree from George Washington University in 1981, and his J.D. law degree, cum laude, from Georgetown Law School in 1984.

Acknowledgement: *I would like to acknowledge Sam Valtenbergs (Associate, LRG Capital Group) with helping me write this chapter.*

A Common Sense Approach to Investing in Biotechnology

Brian H. Dovey

Partner

Domain Associates LLC

Successful Biotechnology Investing

Time and timing are important variables in biotechnology venture capital but they can also be drawbacks.

In biotechnology, once you find a biological mechanism or you have insight into the mechanism, the length of time from that insight to the resulting product is likely a twelve-year process. Our firm tries to look at later stage applications and later stage products, with the intention of forming new companies with them. For example, we have set up new U.S. companies with products that are already on the market in Japan. A number of medium-sized pharmaceutical companies there do not have access to the U.S., so we have started companies using those products. In this manner, we have a company not just with current marketed products, but also a rich pipeline with some home runs in it. In this way, we are able to bypass the long wait from inspiration to production.

There are a number of ways to spot the opportunity or create the opportunity for making an investment in the biotech industry. One way is to develop a name and reputation in the industry, so that people will approach you with opportunities. We have probably the oldest funds, having started in 1985, and we receive around 2,000 deal offers every year. This is a result of our history, including a strong network of entrepreneurs who may be on their third or fourth company; we are also well known to both Wall Street and pharmaceutical people.

The other way, which I think is just as important, if not more important, is to move affirmatively. Constantly be on the lookout for new ideas and developments. I would say 70 to 80 percent of our current ventures have resulted from this. We have gone to Japan and India as well as research institutions in the U.S., to try to monitor what is going on there, and then have started new companies with these new technologies. I do not think it is wise to ever sit back and depend on your deal flow. On one hand, it is good to have the reputation, but on the other hand, you have to be out affirmatively looking and finding new things.

Our approach to investing in biotechnology is different from most of the firms active in the industry for several reasons. During the Internet "bubble," a number of companies pulled back from life sciences or pulled out altogether. They are coming back in now, and tend to want later stage companies—companies that already have sales that are established. I call it getting back into health care with training wheels, because the risk is often lower. We do not do that – in our view, going into those late stage companies in the late stage rounds means lower opportunity along with lower risk. A few firms still do very early biotech deals at the science level, and we do not avoid those, but we will only do a very early biotechnology deal that is a big idea. For example, if someone came up with a new way to treat Alzheimer's disease, or something else of that magnitude, even though the stage is early, we would do it. As a general rule, we will not do early stage investing with modest-sized markets.

Growth and Risk

When we go into a deal, we are looking at trying to generate a multiple of ten times our investment. I think if we can end up overall with three to four times our investment, the investment would be very successful. So at the outset, we look at something that has a ten times potential and we realize it will probably not yield; you will have failures and sideways drifters, so if you can do three or four times, that would be great.

I think many people make the mistake of mixing risk and opportunity together with a thought process that says, if you take an opportunity that is not quite as big, it may be more manageable and the risk may be less. I think we find that not to be the case in biotech. Here, the big ideas and small ideas carry almost the same amount of risk, so you might as well do the big ideas.

How do we decide which risks are worth taking? And, if we do take these risks, how do we handle them? First of all, we perform a great deal of due diligence on every level of a deal. Each deal has its major issues and we will try to identify the two or three drivers. This may be intellectual property or safety or efficacy, but we will spend a lot of time on due diligence trying to understand what the risks are. I think our approach in creating the companies enables us to spend more time on defining the opportunity and understanding the risks. When an entrepreneur comes in with a PowerPoint presentation and says, "By the way, I am on the road this week to see five other firms, so we need a decision in a couple of weeks," it is very difficult to really understand everything that is involved.

In our situation, we might start by funding a company with hundreds of thousands of dollars, rather than millions. We will not put millions in until two or three years out, so we have a number of things going on simultaneously. So when we see a new idea and are smitten by it, we try to put the first $500,000 to $1 million on proving why it is a bad idea. The second best thing to having a success is to have a fast failure in a short amount of time and with a low amount of money.

The second way we mitigate risk is by doing things round by round, so we will put in a certain amount of money to move the company to the next milestone. We are careful to work out with the entrepreneur that if a certain milestone appears to be futile, or the product just is not working - then it is not worth anybody's time to continue to fund it.

I see the biotechnology industry as a rich source for venture capitalists, for a number of reasons. First, biotechnology is really a tool or an engine that is generates many potential applications. This is analogous to the space program, where a lot of the electronics developed for it then spawned additional applications. That is where the venture capitalist really comes in. Doing whole new things in biotechnology is more for academia and much too daunting, but once a modality has been developed, applying that across a number of different areas is right up our alley.

The other thing about biotechnology is that if you have a lot of money and a lot of size and a lot of people, you have no advantage over the start-up. In fact, I would argue this could probably be a disadvantage; scale does not really play a role in applying biotechnology to new applications, so it fits in with the venture capital model.

Beginning Negotiations

To me, because you are about to embark on a long relationship, the negotiations are more about seeing compatibility than defining success, failure, and priorities. When entrepreneurs start talking about how long they can last on the money the deal will bring them, I get concerned. They should be thinking about what can be achieved if things go well using this amount of money. And, if they do not go well, what represents futility, and when exactly it is time to move on.

I ask people if they are interested in fame or fortune—if they are interested in fame, then I do not want to back them. We are rate of return investors and we want congruity in our philosophies. My view on the shaping of the deal is to make sure it is fair from all sides, and to prepare for problems, knowing we have a common interest in fixing them. I think this is particularly applicable when financing is done in future rounds. I stay away from deals with complex financial terms. The more complex the deal is, the less interested I am in getting into it, simply because it sets up different interests. For instance, if someone comes in and they have a three times liquidation preference and the initial investors do not, they would be motivated to accept a three times offer while the initial investors would be motivated to oppose such an offer as they would receive none of the proceeds. Venture-backed companies are inherently risky so it is important that everyone around the table has the same economic stake and the same motivations to help the companies to succeed.

Getting the right CEO is critical to the success of the deal. The "right" CEO may mean a number of CEOs over time. In the beginning of a biotech company, the CEO should be a founding scientist with a lot of fire and desire to move the project along. However, you need to have a discussion with that CEO early on that there will come a point in time, when you are getting near commercialization and clinical trials, that people with different backgrounds and experiences will be needed to move the company forward. That founder needs to be encouraged to see himself as a stockholder rather than an executive. I think as companies move through stages, you frequently need to change the top management team; where we have done that we have been quite successful. The due diligence process can determine bad science, but you never really know good science until you get into humans and clinical trials. You have to do the best you can to vet the risks and look at the opportunity, and once you are in it, the number one criterion for success is the right CEO. In the beginning, the CEO may be a scientist, but as the company matures, you may want to bring in somebody from industry to become the CEO, and that founding scientist then can become CTO, staying with the company but in a reduced role.

After that, getting together an appropriate team of investors and other venture firms to constitute the board is critical, and leads to an interesting dynamic. On the one hand, you want people with different skill sets to be involved. For example, I came from the operations side and have run a pharmaceutical company and start-up, but I do not have a scientific background, so I would look in my syndicate to bring in other venture firms with strong skill sets in the scientific area. On the other hand, you want to have people with compatible philosophies, who look at risk and do not panic when bad things happen, who are more oriented to fix things, and who are not looking for quick exits but want to build really good companies. We do a lot of ventures on our own, but we have also done syndicates and formed a lot of specialty pharmaceutical companies, and there we had people with different points of view but the same philosophy on company building.

Challenges in Negotiations

What the different sides in a negotiation want depends on where the deal is coming from. For instance, if we are in-licensing products from another country like Japan or Germany, we pretty much understand what they need. Normally, they are as interested as we are in an equity stake, much more so than royalties or cash. Still, we are happy to do those as well.

I think that negotiating the right deal is grossly overstated. In early stage biotech venture capital, I know of no situation where having done a really sharp deal yields great returns, and having done a bad deal causes a drag down in returns. Because your relationship is going to be a long-term one, say five to ten years, if any venture people try to take advantage of the entrepreneur, he or she will eventually figure it out. Thus, doing a bad deal with the entrepreneur is not a good way of doing business. Similarly, we stay away from entrepreneurs who have unrealistic views of what the company is worth. To take the entire process from where we see a deal to where we consummate one, the amount of time spent on negotiating the financial terms is probably 5 percent, and there are usually a lot of comparables around so that everybody knows what zip code they are going to be in.

I think probably the thing you spend the most time on is trying to get agreement as to the basis of continuing to fund it. This is not saying, "If you do not hit this milestone, we are not going to fund you anymore," but more of trying to define futility. For example, we had one deal in Alzheimer's that did not work out. We said we would invest $20 million, but if, in fact, after X number of patients we are not seeing some split between the control group and the treated group, then, even though this treatment may be effective, it is only very mildly effective and unlikely to be embraced by the medical community. We spent a good amount of time deciding how many patients it would take to determine what I call futility, and we came to agreement on that. It was not in our interest or the CEO's interest to work on and fund something that is unlikely to be valuable in the future. As I said to the CEO at a point, "It is wasting our money, but more importantly, it is wasting your life to continue on if it is futile."

Time is spent on what the first round of financing is trying to achieve, and then mapping out what the subsequent rounds would look like. Then follows the discussion of what the initial stake should be and what the pre-money is and the option pool, but as I said, there are a lot of comparables around, so we pretty much come to an agreement on that.

The most difficult factor in a deal is agreeing on what the founder role is going to be going forth. You can hold that discussion with a founder, and because you are the one financing it and he is the one looking for the financing, it is easy to agree in theory to some kind of succession over time. As a practical matter, it usually becomes more difficult or problematic at the time that you think it is appropriate to make a switch. They may have agreed with you in principle at the inception that a switch should be made, but now are saying, "Gee, not now, maybe in a couple of years from now." So you have that dynamic going on, but it really is trying to get a good understanding of what everybody's role is going forward.

The way to handle that situation depends on the individual involved. In my opinion, addressing the issue right at the beginning is the right thing to do. There are other venture people who think, "When the time comes, we will all know it, and I do not want to risk losing this deal by telling the entrepreneur that we may bring someone else in there." I actually do not share that view, and again, think it is better to have that discussion with the entrepreneur at the beginning. If they say their heart has always been set on running a company, and Bill Gates has done it and that is what they want to do, I am unlikely to invest.

There are many other reasons to walk away from an otherwise good deal. I think sometimes you see people with unrealistic expectations as to what something is worth. We looked at a deal in Europe where I thought it was a great idea. Comparable kinds of ideas at that stage had a pre-money value of $4 to 5 million, and this person thought this idea was worth $100 million. That thinking is going nowhere, because if you are at $4 million and they are at $100 million, then compromising at $7 or $8 million is never going to make that person happy, so we let them get experience elsewhere.

At this point, we are seeing literally thousands of deals, looking hard at between one hundred and 150 and doing ten. There are many things that can go wrong, and there is no decision criterion or algorithm that you can put over a business plan to determine good ones from bad ones, and there are no overriding comments that can be made in this regard. Sometimes people will say unless you have a great, experienced CEO, it is not going to work, but I have seen some inexperienced CEOs where the idea is so powerful that you take that chance, and vice versa.

The context of all these factors and variables, coupled with the ability to think about all of these pieces of information and integrate them into a decision, make this job a challenge.

I think the best advice I can offer anyone in this business is that you need to be a very heavy listener. If you have a great amount of experience in a particular area, that can be detrimental. You need to subvert your ego, and everything else, in order to really talk to the experts that are relevant to the particular opportunity you are looking at, and not let your own views get in the way.

Experience obviously cuts both ways—it is a good thing and a bad thing. Just because you have seen a couple of failures in a specific area does not mean that area is not worth doing. You need to have an open and flexible mind. Indeed, the real talent here is not simply bringing your expertise to decide on what is a good or bad thing, it is how you marshal an effort to find out from the real experts what is good or bad about the deal and then differentiating out where there are problems or risks. You must ask if they are risks you should not invest in, or risks you should invest in—and then try to vet during the company formation process or first year of operation.

The skills in venture capital are less technical in terms of having Ph.D.s or M.D.s, but more generalist. You must know the area, but the ability to integrate the information and make a sound judgment is crucial. We have found in our firm that 75 percent of our successes are actually not practicing the business plan we invested in, so you need to be able to accommodate a lot of mid-course corrections.

Advice for the Management Team or Entrepreneur

These people investing in your product are going to be with you for the long term. You want to make sure that they remain with you, and you are going to need their continued support as you move ahead round by round. A one-time investor who does not intend to invest in subsequent rounds is going to be a problem, because a new investor is going to assume that since they are not participating, they have found the fatal flaw. I think it is important to pay attention to the quality of the firm that you are working with and the people who are going to be involved.

I think entrepreneurs are somewhat reticent to do due diligence, and I think that is a mistake. Entrepreneurs should find out what other companies the firm has funded, call the CEOs of those companies, and find out what it has been like to live with the firm.

Many entrepreneurs will want to get the best financial deal, and I would discourage people from doing that, because I think that if you go around enough, you can always find someone who will overpay. This is fine, except you want to be sure that you do not create a problem. I have seen deals where the initial round is doctors defined by maybe fifty having invested in it at a very high price and creating a situation such that the post money value is $10 million and the company is probably worth $3 to $4 million. We will not look at those, as I have no interest in confronting fifty doctors and explaining to them that they grossly overpaid in the beginning, so we would just pass on that situation. This is a shame, because the entrepreneur can set a financial structure that can be to his or her detriment.

Sometimes this problem is related to capacity. The doctors who funded the venture in the first round did not have the capability to fund it the next time around. Sometimes a small venture firm, unless they have a relationship with larger venture firms, may be in this situation. Similarly, there are some firms that say, "If you do not hit your milestones, we will not fund you anymore." If a firm has that kind of track record, I would tell the entrepreneur to be very careful. All entrepreneurs go in with a spirit of optimism, but the better question to ask is if a company did not hit the milestones, but did the right things and learned a lot from it, did the venture firm continue to fund them? If the answer is yes, then that is the kind of firm that I would want to deal with, not one that automatically cuts me off if I do not hit the sales target of X.

Also, entrepreneurs should not just send out business plans to as many people as they can. You need to really network into the venture firm. You should independently try to sit back and determine who are the top five firms you want to deal with, and then figure out how to get to some of the top people at that firm through your lawyer or accountant, or other venture people or entrepreneurs. When I get a plan that says "Copy #1040, Highly Confidential," that comes over the transom, we do not even read it.

Valuations at Different Stages of Investment

Right at the beginning, a biotechnology company's valuation is in the $2 to $5 million range, if you have what looks like a promising idea with a product maybe a year off or so. By product, I mean something where they have an insight on a biological activity and an idea as to how a chemical might block that activity. For instance, anti-cancer is probably in the $2 to $5 million category.

Something that has shown efficacy in animals and safety and is about to go into human clinical trials is probably going to be more in the $10 to $15 million range, and if it has gone through Phase II clinical trials, in other words showing it is efficacious and effective in a small group and has been through the safety, you are up to $25 million or so.

All these valuation stages are very dependent on how big the opportunity is, too. If you are dealing with a big disease or small disease, there is a lot of variability. If it is something that has gone all the way through a drug approval and they are about to set the stage for marketing it, usually you might be able to do an initial public offering at that stage in the $100 to $150 million category.

Establishing Valuations

As a practical matter, establishing valuations really is predicated on comparables. I teach part time at an M.B.A. program and many students are interested in this. I give them a case in which there is a fairly complicated Monte Carlo analysis with discounted cash flows, etc., and I end the class by saying there are not really any venture capitalists who do this. It is all determined by looking at other valuations of other companies.

If everything works according to the way they are saying it, there should be a possibility of a ten times return or something in there and if everything goes right we will wind up making 15 percent—it makes no sense for us to go in at that kind of valuation.

Again, as a practical matter, establishing valuations is heavily predicated on comparables. But, as negotiating matter, it is one of having a very open process with the entrepreneur. If they ask me what kind of returns I am looking for, I will tell them. I think it needs to be quite transparent and quite open and I encourage them if they have a completely different view of it to talk to others. Do not go into this taking our money with an attitude like, "Well, I will take it because they are offering it, but I am getting a bad deal on the valuation." If you can do a better deal, do it.

I do not think I ever really know for certain if I did a good job establishing a valuation. If you get a company start-up, and you are going in with a valuation of, say, in the $3 million category and it is a success, if I look back and say if I had done it at $2 million how much more successful it would be, or at $4 million how much less successful, I would find the difference diminishes. So, at the early stage we are not necessarily that valuation sensitive within a range, we are talking in the single digit millions. We do not go into any of those hoping that the company will be worth $15 to $20 million, we all go into it thinking the company will be worth $100 million-plus after a couple of rounds, so as I said, I do not think we are that valuation sensitive.

I think for those playing in later stage deals where the valuations are up in the $100 to $150 million, they are very valuation sensitive, and you really need to look hard at what kind of return you may or may not be able to generate at that level.

Factoring the hopeful rate of return into a valuation is, in my way of thinking, less of a calculation and more of a negative. If in fact somebody has a valuation view that is much too high, we will just not do it. With brand new science in start-ups, we are not trying to calculate a rate of return, as there are too many variables. Say somebody was going at a cure for Alzheimer's, if it works just a little for a subset of patients, it is probably worth billions of dollars. If it works more generally, it is billions and billions.

The other issue affecting the hopeful rate of return is the market size. When you are talking about something like Alzheimer's, the market size almost does not make any difference because it is such a big number. You then need to focus on the likelihood that this will really work, and if it does not work, how fast can I find that out?

We look at valuation much more in a "Yup/No" manner. If the entrepreneurs clearly have delusions of grandeur, we are not going to do it, but if they are generally in the same ballpark, we will compromise on how to work it out.

Other than preferring that all sides really see eye to eye on the valuation, to me the most important thing is that the entrepreneurs do their homework, talk to others, and get themselves comfortable with knowing that what they are getting is fair and equitable.

We are seeing more and more entrepreneurs do this for the second and third time now, and they know exactly what they are doing. I think those entrepreneurs who are new to it will talk to other entrepreneurs when they calculate their valuations, as well as to their lawyers and accountants who are experienced in this area. It becomes a fairly easy process when they do that. The ones who are difficult to deal with when negotiating valuations are the scientists who believe they have the answer to all the world's problems, and that their situation is extremely unique and valuable. When they come from a perspective of naiveté is when we have a problem. If they get the best law and accounting firms that are really experienced in this helping them in the valuation, I find that is a positive thing. You can get professionals who know what they are doing to come to a consensus. It is when they are amateur entrepreneurs that you run into problems.

When trying to establish a higher valuation, you need to think about what the end is going to be. My advice would be to see what the venture capitalist thinks about whether the company is going to get sold, or at what point an initial public offering would be possible, and what kind of values are there, then start walking backwards. Assume I show this and it is effective in animals and we go to the next round. What do you think the price of that will look like? The entrepreneur's focus is usually on getting that initial value right. He needs to understand exactly what the process will look like going forward and what is going to happen if we hit some bumps in the road. Is the venture capital firm going to take that out of them in the valuations or will the firm support them if they have a problem and do not get there? Will the firm support it at the last round price? Then the entrepreneur needs to go out and verify the track record of that venture capital firm on what they have done with other companies they have funded. After having done all that homework, if everything adds up, then approach the venture capital firm with the higher valuation.

In Conclusion

Some people say there is too much money chasing too few deals and from the entrepreneur's perspective sometimes they are just not funded enough. I think neither is true. I think the opportunities we are seeing are terrific and I do not think there is too much money chasing too few deals on the other side. I think the things that are not getting funded, which is unfortunate, are very early stage science. In the early 90s, we would fund things right out of the university, but that is just not feasible anymore, thus many good ideas are not getting funded. My thought is that we as a country ought to figure out another way to do this, possibly with government help.

Brian Dovey has been a partner of Domain since 1988. Since joining Domain, he has served as chairman of Athena Neurosciences, Creative BioMolecules, and Univax Biologics and on the board of directors of more than thirty additional companies. Present board memberships include Align Technology, BAS Medical, BioVascular, Neose Technologies, Ocera Therapeutics, Orexigen Therapeutics, Orqis Medical, REVA Medical, SkinMedica, Symphony Medical, and Valley Forge Pharmaceuticals. Past board memberships include British Biotech, Cardiac Science, Connetics, Geron, Ista Pharmaceuticals, Nabi, ReSound, Trimeris, U.S. Bioscience, and Vivus. Prior to Domain, Mr. Dovey spent six years at Rorer Group (now Sanofi Aventis as a result of subsequent mergers).

As president of Rorer from 1986 to 1988, he was the primary architect of this Fortune 500 company's strategic shift to pharmaceuticals that resulted in a doubling of annual sales to approximately $1 billion. Previously, Mr. Dovey was president of Survival Technology, a start-up medical products company whose sales growth placed it in the top ten of the Inc 100. He has also held management positions with Howmedica, Howmet Corp., and New York Telephone. Mr. Dovey has served as both president and chairman of the National Venture Capital Association. He is chair of the Wistar Institute and a former member of the Burnham Institute, both of which are leaders in preclinical biomedical research in the non-profit sector. Mr. Dovey also serves on the board of directors of the Kauffman Fellows Program at the Center for Venture Education. He is Trustee Emeritus of Germantown Academy and a former Trustee of the University of Pennsylvania School of Nursing. He is also a former board member of the Advanced Medical Technology Association, which represents the medical device industry and the Consumer Healthcare Products Association, which represents the non-prescription pharmaceutical industry.

Mr. Dovey received his B.A. from Colgate University and his M.B.A. from Harvard Business School.

Taking a Product From Concept to Approval

Harry Ross, M.D.

General Partner

Aweida Capital Management

I used to be enamored with new technologies and broad platforms. The far-ranging potentials of a new science made it difficult to settle down and commit to a single project. This often led to lack of focus within companies as it's tempting to jump from project to project but never finish one completely. It is very easy to become captivated by a technology in search of an application. More important though is to intelligently choose and pursue a single-defined application. One of the hardest challenges for a new company can be defining the first product and committing to its full development before chasing the next application.

In today's market, investors are generally less interested in broad concept stage platforms. As a venture capitalist, I want to see the finer points of the specific product and the method by which it will be created. Beyond the details of exactly what drug, device or application is in development, it's distracting to hear all the unlimited possibilities of what could be done with unrestrained resources. More valuable is to understand in clear concise terms how a lead product will progress from concept through approval. Second and third products are important for a healthy pipeline but these are discounted by orders of magnitude after the lead candidate.

The key to achieving overall success in today's market is to maintain focus on moving a defined product from concept to approval. Ultimately, the ability to achieve clear milestones on a single project in reasonable time frames is a great value driver.

Seeking Out Opportunities: Taking Initiative

Beyond being just a source of cash, I have a great interest in both developing technologies and founding of new companies. There is a great sense of accomplishment in actively participating in these early stages. The fine line in this process is understanding when to engage and when to step back. In the life of a company, it is not unusual for the value of an individual's contribution to change. Mine is no different and keeping a perspective on this is critical.

I look for opportunities in the industry where I see an obvious hole in the current state of the art. In doing so, I try to understand why a new concept will not only improve upon the current standards, but will actually be embraced by the people who must use it. Weighing the realistic needs of clinicians versus exciting but impractical science projects can be quite difficult. There is a delicate balance between improvement and expense, as the new product must provide sufficient improvement and inducement for patients to abandon their current therapies for various levels of gain.

I am most interested in seeing products that present a leap forward in terms of being cheaper, faster, better, and easier and showing more potency or safety than their predecessors. A slight improvement is not enough to get me excited about a deal. For better or worse, nearly every disease has a current gold standard of treatment, and it is vital that new products far surpass that standard.

The Scientific Challenges of Biotech VC

One of the more difficult aspects of health sciences venture capital compared to other sectors is that it involves not only rigorous engineering but also often unpredictable science. Results are difficult to predict, and the outcomes are not always expected or anticipated.

I often look at software or hardware companies outside of the health sciences, and in many cases I find that the problems I see are matters of engineering. With enough sets of hands and brain matter put to the task, I know they will succeed at reaching their goal. Unfortunately, playing with technology in biologic organisms involves a number of unknowns, and no matter how many minds are put to the task, the desired outcome may not be achieved.

Evaluating Returns

If I do not see the opportunity for at least a ten times growth, I typically don't consider the company worth the risk. Much of this evaluation is based upon market size. Fortunately in health sciences, many of the markets are already well defined. For example, if someone approaches me with an idea for a new drug or device, I am able to examine the prevalence rates of that specific disease. Current treatment modalities can be quantified by revenues and market share. This is a relatively quick and simple process, and an easy way to judge the value of a potential investment.

In some ways, it is easier for me to understand market needs in health sciences versus other sectors, such as software or high tech. Most diseases have been well studied and defined over time. The current state of the art may be inadequate, but there is a general understanding of needs and size with a baseline to work from. In contrast, I often come across sectors with applications for a previously untapped market. It can be difficult for me to gauge the acceptance of an innovative technology in an area that has few previous benchmarks. Assuming a successful development process, predicting adoption can be challenging.

It is also important to look at how disruptive the new technology will be, how big a dent in the market it will impact, what the costs and reimbursements will be, and which way the trends are moving in the health care industry at large. Even after the potential market is projected, there is no shortage of variables that must be considered

Technology and Risk

Technology is a large part of the risk equation in the biotech field. There are always questions regarding whether the product can be synthesized, whether it can pass through the regulatory and approval process, and whether there is a true market for it. However, above and beyond these technological concerns is the risk associated with the execution. Is the right team in place with the proper skill set for the task? How does the execution compare to the competition? How likely is it that a successful end product will be unique in the marketplace? These questions are the most difficult to answer, but must be factored into the risk equation just the same.

The Management Team: Balancing Experience with Integrity

I look most for integrity and experience in the management team of any company I am considering. Experience is a wonderful thing to have, but it is important to remember that no matter how experienced a team may be, without integrity the risk is too high. Furthermore, once a team/company is approved for financing, the individuals will be responsible for wearing a dozen different hats, and it is rare to find anyone who has prior experience with all of them. There is no such thing as the perfect team. One looks for the best possible combination of technical, financial, and clinical experience that will enable the group to best handle the tasks at hand.

At times management teams run into trouble by indulging in overconfidence. It can be tempting to underestimate the difficulty of the task, the size of the risk, technological and regulatory hurdles, or the overall challenges of moving from point A to point B in the growth process. Knowing how to handle setbacks and detours along the way is a critical skill set because these unexpected setbacks are inevitable.

The Test of Success

The true indicator of success is that people are eager to work with you again on the next deal, regardless of whether they are investors or founders. Sometimes one does everything wrong and the company finds a successful exit, and other times one does everything right and the company ultimately goes under. For this reason, I judge my own success by the number of calls I get from people I have worked with in the past who are interested in pursuing the next deal. Repeat business means that I did my best to put things in their right place, to treat everyone fairly and honestly, and to offer some value beyond merely the checks that were written.

Deal Terms for Biotech Ventures

The most important thing on a venture capital term sheet is that expectations and contingencies are laid out clearly for all to see. A recurring problem with term sheets is that concepts are not clearly defined, which is often a setup for failure. If I am expecting A and the other side is expecting B, neither of us will be happy.

When I structure a deal, transparency is a top priority. I want no surprises whatsoever. Surprises at this stage often create animosity that is difficult to resolve or an outright deal killer. I do my best to set up the term sheet such that the investors and management are aligned. This prevents an "us versus them" mentality from creeping into the deal. Ultimately, both sides of the deal must work together, and an ugly or ill-defined section of the term sheet may ruin the trust that is necessary for the company to succeed.

To this end, I also set up protections for both sides such that in the event that problems arise, there is a clear path to resolution. Equally important as the deal terms themselves are the stipulations for how questions or concerns will be handled. Addressing each of the "what-ifs" beforehand is the best way to mitigate problems down the road.

All investors have a rough template that they are comfortable working with. This includes those areas that have worked well in the past or protections against prior problems. In spite of these established methods, we are willing to deviate from normal deal structure in the case of entrepreneurs with whom we have previously worked successfully. If a tried and true entrepreneur believes that an atypical approach will ultimately benefit the company's success, we are more willing to step out on a limb.

Valuation and Other Structural Concerns

Once everyone has agreed to align themselves and the encompassing language is in place, the next level is valuation. There are no shortage of tools and algorithms out there to help establish this. When I look at valuations, I am most concerned with the stage of the technology, when it will come to fruition, its needed skill sets/resources, and the associated risk. I also consider the science behind the product, size of the market, and the value to potential patients. Entering in at the early pre-revenue development stage means that one need not discount earnings or work on MPVs just yet, focusing instead on the sheer likelihood of success. An average concept stage early start-up company might have valuations measured in hundreds of thousands. One that is in its second year pre-revenue might come out of the valuation process in the $2 to $5 million range, whereas a later stage but still pre-approval company might look at $10 million plus.

Additionally, the parties must decide what the cap table will look like and how much money is necessary to achieve success. There must be appropriate holdbacks in place for employees and stock pools must be well defined. As an investor, I want to be absolutely certain that the employees are taken care of. Disgruntled, unmotivated employees can destroy the best technology quickly. Employee stock plans create the sense of ownership that is required for the dynamic creative process necessary for success.

One needs to look closely at intellectual property. This is not just patents but, equally important, freedom to operate. Ideally, one should encounter no roadblocks in the company's ability to do exactly what it has set out to do while also preventing competition from obtaining access. One must consider at what stage the current patents are, how broad they are, and what other patents might exist in similar areas. To this end, it can be helpful to set informal milestones in order to keep a handle on the project's progress. Patent work is an ongoing process that starts upon founding of the company and continues through its exit. Strong positions not only prevent others from capitalizing on a company's proprietary work but also assure the freedom to operate in one's stated field.

Other concerns that arise in structuring deals will depend upon the occasion. One must consider whether there are licenses to be dealt with, whether there are rights or royalties owed, whether the technology has been split, and so on. Any number of these variables can impact a deal, and it is critical to be aware of them from the beginning before attempting to determine where the company is ultimately headed.

Negotiating with Founders

There is typically a good deal of negotiation room in working with founders and entrepreneurs. Minor factors crop up in great numbers and are typically dispensed with easily. Major factors, however, can lead to problems and in some cases may kill a deal outright. Examples of deal killers might include stock structure, property rights for future inventions, commitments to work solely on the project, or the amount of resources that will be devoted to the undertaking. The finding of last minute surprises or revelations is one of the most difficult problems to overcome. As previously mentioned, the perception of trust is paramount and last minute revelations are difficult to overcome.

From the founders' side, overvaluing early concept stage technology is the most frequent source of friction. If the concept is not valued on its current stage combined with the likelihood of success, disagreement will arise. It can be tempting to value a project based solely on the size of the market with little attention paid to the project's current stage of development or chances of success. In this setting, milestone structures can be very helpful. If the project goes smooth and quickly, it may deserve a higher valuation. If there are delays or setbacks, a lower valuation may be more appropriate.

Late Stage versus Early Stage Investments

In working with a late stage established company, much of the deal structure is already in place. As an investor in this situation, one is typically working with only slight modifications on a well-established entity. Prior investors and management often have constructed a working unit that is meant to grow and not necessarily be disrupted by an upstart new investor. Smaller or later investors may be given a more limited choice of options from which to choose. Ideally, they are joining a well-established body that does not need or want disruptive changes. In dysfunctional companies and depending on the relative size of the investment, change may be expected.

With early stage investing, one is in the position to create a new entity on a fresh slate. With minimal or no form in place, the company can be molded into an organization that works well for the given needs. There is rarely a single rigid structure that is appropriate for all situations. At the early stages, the specific requirements of the company's execution can be evaluated and built in accordingly.

Working with Other Investors

I usually work with at least one or two other investors at any given time. Beyond that, there are simply too many cooks and it becomes increasingly difficult to perform effectively. It is often the case that no matter how many groups are actually writing checks, only one or two are actively driving the process. Having a core group in place can streamline the process of presenting options to the assembly.

Most investors tend to end up working with others who have similar strategies and standards. I do my best to work with investors who think along the same lines as me, as having similar approaches simplifies the process. At the same time, however, I seek out smart engaging groups that will challenge us on the various points of deal structure and execution.

We do not often alter deal structures based on competition. Either we decide we are going to work together with an entrepreneur or we pass. Entering a bidding war is a sure indication that there is either not enough interest on my part or theirs. On the other hand, deals are often lengthy processes, and new information that arises over the course of the negotiations may impact valuations and risk. Resources required to execute the project may go up or down. Key individuals may be added or lost. Competition may change. All of these need to be taken into account and evaluated.

Harry Ross, M.D., is a graduate of Oregon Health Science University. He received post-graduate training at Hennepin County Medical Center in surgery and emergency medicine with subsequent Board Certification in emergency medicine. Now retired from clinical practice, he is a managing partner in Aweida Capital Management where he specializes in life science investments.

Strategies for Biotech Venture Investments in Early Stage Companies

Dr. Nelson Teng

Managing Director

Charter Life Sciences

Where We Came from and Where We're Headed: History of Biotech and Vision for the Industry's Future

I have been in the biotech industry for the last twenty-five years. Time has changed the industry, and it has matured into an important sector of the economy. In fact, its evolution has become a very important part of the technology development in the United States.

In the past twenty years or so the biotech life cycle for the investor, although treacherous, has as a whole been very successful. In fact, the return for life sciences investments has performed better than the tech and IT sector according to venture investing data from the past twenty years. During this same period, early stage biotech VCs outperformed later stage biotech VCs, and both outperformed the balanced VC fund. Going forward I believe early stage investors will continue to harvest the best returns. However, as the industry evolves and matures, I anticipate there will be a stage migration. For example, a Series A investment ten years ago would now be considered a seed investment.

For the first wave of biotech, you have to give credit to two major developments: the recombinant DNA technology and the development of monoclonal antibodies. The second wave of biotech is attributed to the technology that enabled the rapid sequencing of DNA, RNA, and proteins, coupled with the ability of vast information gathering and processing—the so-called genomics and proteomics era. In addition, new technology enabled novel methods for making chemical compounds. The biotech industry was further stimulated by new understandings in biologics and receptors, and the discovery of the many signal transduction pathways. The third wave of biotech investing materialized with the advent of micro RNA, nanotechnology, and stem cells - I would characterize this last wave as the clinical trial wave.

Up to the mid 1990s or so, VC investors have been focused on platform technology and company pipelines. When the biotech industry first started, any professor coming out of Stanford or Harvard could take a couple of pieces of transparency, go to the investors, get funding, and continue to do their research in a new environment with more money and support. At one time not long ago, we used to joke that if a company started to hire a lot of

M.D.s, then that was the company you should invest in - because it meant they were getting close to an end product. After the IT bubble crash and some of the crashes in biotech, a company must demonstrate that it has a product in order to obtain investment. This shift in focus has concurrently created an increase in the number of clinical trials being conducted, as well as the number of biologics in clinical trials.

The old business model of biotech companies forming from start-up venture funding and growing organically into significant companies that may eventually go public is over. In other words, the era of Genentech-like companies is nearing its end. Few companies growing organically can make it in today's environment for various reasons, including the fact that most venture funds are not supporting companies through a ten to fifteen year growth cycle—they are looking for a more immediate exit. To really demonstrate that this era is over, even Genentech itself is now part of Roche. In a sense, the biotech companies become the research houses of the pharma. Investors need to be cognizant of the evolution and the change in the business model and structure in the biotech industry to be able to adjust their investment strategy accordingly.

My vision of the future is twofold. The biotech industry has reached a stage at which a company cannot solely be a research house. If a company has a platform technology, then they will have to demonstrate its feasibility and translate that into commercialization. They also must demonstrate the ability to go through the regulatory, clinical trial, and manufacturing challenges. As the industry has matured, there is an increased focus on the potential success of these companies.

In the automobile industry at the turn of the century, there were many automobile makers. Later, these were consolidated, and now we are left with just a few giants. Given that biopharmaceutical is by definition a worldwide industry and it takes a great deal of horsepower to drive it, unless you are a specialty oriented boutique house in a diagnostics area or medical devices, most of the pharma and new drugs will be consolidated into larger pharma. By this I mean that as the biotech industry matures into an important sector of the economy, many of the start-up biotech companies will merge or consolidate into these larger pharma.

Spotting Opportunity for Investment

At Charter Life Sciences ("CLS"), we tend to shy away from the health services and bioinformatics areas, mainly because of our internal skill set and past experience. Aside from disruptive technology—a unique technology which has an overwhelming advantage over existing technology—we look for something more than a pure technology platform. Nowadays, the investor and the company realize that it is not just pure technology that wins. In marketing, understanding the financial structure and establishing a team that can execute are some of the more important factors. Other important components are the speed at which the product goes to market and the company's capital efficiency. These factors can make or break a company.

We look for several things, the first being unique technology which has developed into a translational stage, addresses an unmet significant healthcare need and has a clear pathway to a product. We look for a product that is close to human clinical trials, commonly called first-in-man testing. Another area that we investigate carefully is the type of space that the particular technology occupies. Is it in an area where there are a hundred competitors with similar products? VCs are supposed to make investments in areas that are challenging and risky, but at the same time, most VCs are rather risk adverse. CLS is not risk adverse—instead we just try to be financially smart and management savvy. It is critical for us to put the money in an area that can make a real contribution in technology development to reach a milestone result that produces a value inflection for the company.

The other important factors that enter into our decision-making process include the team experience and dynamics and market factors. You need a professor or an entrepreneur with a great track record in product development, not just in research, and a team that has worked together and has the ability to execute, and a valuation that makes sense when gauged among comparables. We look for how important the product is. Is it going to be a paradigm shift and a real novel innovation? The last thing that we have to be sure of is that the project is realistic and must be within budget and financially sound.

A Different Approach to Biotech Venture Capital Investing

In the early days of biotech investing, a large fund would have one to two life sciences partners in a much larger group of technology partners. As the biotech industry became more sophisticated, the one to two life science partners were not able to get the critical mass needed for adequate due diligence. As a result, many weekly partners' meeting would end up being an asset allocation discussion rather than a discussion about the basic technology of each company, the technical fundamentals or investment criteria. As a result, there has been an exodus of biotech focused partners to form their own biotech specialty funds.

Charter Life Sciences is a clinical stage life sciences focused fund. First off, CLS is not a seed stage investor. We think that is the job of the government granting agents, foundations, and angel investors. We are really more Series A and Series B investors. By definition, we specialize in early stage companies. We are partners who have all worked in the industry, have founded companies, have operating experience, and have gone through the ups and downs of the life cycle of the industry. Our partners include physicians who have practiced, executives who have operated companies, a regulatory expert who has consulted for clinical trials and interacted with the FDA for over twenty years, and a lawyer and CFO who has been in biotech investing for the past twenty-five years. Most of us also have international business experience.

We differentiate ourselves from most other funds because our partner composition is uniquely specialized and we have worldwide ties that connect us to opportunities on a global level. We are able to source deals from all over the world, from Europe and Asia in particular. In practice, it is difficult to be an early stage investor and be global, but with extensive networking it can be done successfully. We also believe that great opportunities exist all over the world. We can't assume that there is only one place on earth that technology can come to and that being Silicon Valley, USA. There are good, strong scientists and strong human clinical data being done in a variety of geographies. The cross border deals are often more difficult to execute due to cultural differences, logistics, distance, and level of understanding of the local operating and regulatory environment. We have the expertise and worldwide connections that enable us to intelligently evaluate global companies and, after investment, to guide the company in its development.

The biotech industry can often scare LPs away with its unpredictable reputation. As a biotech investor, you are at the whim of the FDA, you have a long incubation time and many public biotech companies have yet to turn a profit. Biotech has gone through several evolutions in the last twenty-five years, and is in the maturation phase. This phase is a time to harvest the fruit but it requires specialists like the ones we have on our team to realize that advantage.

Biotech VC Challenges

The biotech field is a constantly evolving and dynamic young industry that can appear to be unpredictable in its evolution. However, for those of us who have been in the industry for the past several decades, we see well defined pathways and regulations for the product development process as dictated by the FDA. In many ways, the FDA process is one of the most challenging hurtles of moving a company forward. Another distinctive feature of the biotech industry, which is both a challenge and an advantage in biotech investing, is that information on your competitors is readily available. FDA trials are public information so you can obtain information on the trial sponsor, the product being tested, the patient population and a number of other details.

The design and development of the clinical trials is a very important and crucial process for the biotech company. One of the earliest hurtles is to obtain an early proof of principle trial in man. Once that is successfully completed, the company will move on to the pivotal human clinical trials. The common understanding is that it is risky to predict the success of a drug even after the completion of a Phase I or a Phase II trial. Many would harp on the fact that products can fail even after advancing to the Phase III stage. We believe that this is a rather selective view of the process that may not account for regulatory experience and good trial design. If you look at the success of clinical trials conducted by large pharmas versus small pharma companies and universities, you can see varying rates of success. To me, that is a reflection of the sophistication and skill set of the people doing the clinical trial design as well as the experience and knowledge of the company team and investors. We consider ourselves to be an evidence-based selector of companies. We don't think it is left as much to chance as an amateur aiming for the bulls eye on a dartboard. We are very selective and strategic during the planning process—the biotech venture investing is really a thinking man's game.

As I have mentioned several times above, the company's team is very important. It is absolutely critical to have the right CEO who can guide and lead the company at that company's particular stage of its development. An experienced team can navigate the troubled biotech water in the event of a clinical trial disaster and can potentially correct the course in a financially sustainable and capital efficient manner to ensure the survival of the company.

Measuring Potential

As early stage investors, we generally don't worry about the market growth potential, as our companies are too early in the process to be influenced by the market. A company's growth potential is generally of concern for late stage investors. At each round of the investing, we evaluate the value inflection potential and the multiple. For instance, from Series A to Series B, we like to see at least a two times appreciation of the initial investment value. Our investment strategy is that each round of the money has to be budgeted sufficiently to reach the next pre-established milestone. We expect the company value to increase at each milestone. Thus, we determine the early stage company's growth potential on its ability and likelihood of success in execution.

Risk Management

All early stage companies carry significant risks with their technology, people, market, and potential for financing. In our review process, we analyze each of these factors independently and manage the risk separately. To address the technology risk, a detailed analysis of the science, pre-clinical data or clinical data if available is a must. People risk is best managed by having the right CEO. For example, the scientific founder frequently is the initial CEO who may lack adequate operational experience. After an investment is made, the VC needs to be able to jump in and to guide the operations of the company. As early stage investors, we focus on reducing the clinical risk as early as feasible. We reduce the risk by looking at how the proof of principle study can be done in a minimal number of patients. During this early process, the risk is managed by a clear-cut, well designed protocol. If the trial reaches its designed clinical efficacy; it should provide sufficient data for a decision for further investment.

Other areas that we scrutinize with a great deal of care are the IP and IP/technology licensing agreements from universities. The IP for biomedicine and medical devices are complex and, at times, confusing because the biological process, unlike a computer processor, is not so clearly defined. Many of the patents may be pioneering and have not been tested legally. A significant future risk can be reduced by a strong IP position and a clean licensing agreement.

Every company comes with a financial risk and we manage this risk through a syndication of VC investors. For early stage company, a syndicate of investors ensures adequate funding and provides the experience and guidance to manage the team, company structure, and product development planning. Proper management of an early stage company is critical and is a must in managing the risks and increasing the chances for success.

Common Mistakes

The most common problem I have seen on the company side are founders who underestimate the risks of their venture and believe they have the skill set to successfully move the company forward in its various stages . In some cases, the original founder may have the skills and experience to run the company during the early clinical stage of the company. More frequently, an experienced CEO will need to be brought into the company to move the company forward after the initial proof-of-principle studies. The reality is that good science and good technology can be buried underneath the weight of a sub-par management team.

A common problem I have seen on the venture capital side is that large VCs can lose sight of an early stage company's value proposition. Large VCs are constrained in their investment activities by their large fund size. Their fund size constrains their ability to invest their capital in the most capital efficiently matter. Additionally, VCs with only a financial background can mis-guide the company's technology development by insisting on certain direction or strategy.

A common problem for both entrepreneurs and VCs is when they have opposing perceptions of the company's technology value, financial risks and business development needs. Unfortunately, this difference in opinion can result in significant conflicts, stress and management challenge.

Achieving Long-Term Success

There is no single right answer to the question of how to achieve success. Success is the result of a complex combination of sound financial fundamentals, a good technology that generates a product that fills an unmet medical need. The right combination of these elements can produce a recipe for the long-term success. Because of this, my feeling is that in the early stage of the biotech sector, you have to be both technologically aware and financially savvy. Also, on a more basic level, you must have people skills to do early stage investing. The successful building of companies over long term should translate into long term success of a fund.

The Structure of a VC Deal

Our firm's standard term sheet covers the following items:

Pre-money valuation
Size of round
Investors
Dividends
Liquidation preference
Conversion right
Redemption
Anti-dilution
Voting rights
Investor's rights agreement
Board of directors and voting agreement

Most VC deals are structured in a fairly standard way. The details of the terms depends on which funding stage you are at and whether you have a general agreement with the entrepreneurs. Our investments are usually structured with a syndicated group of VCs that might include industry partners. The standard deal would have an equity portion for the founders and reserve an option pool for the employees. Depending on the agreed upon milestones and future employee hires, additional investment amounts can be tranched. In many cases, certain preferred rights and liquidation preferences for the VCs may also be included. The pre-money valuation is the most frequently negotiated topic at the early stage investment level.

The most important thing for the VC to do in structuring a deal is to have a detailed outline of how the proceeds are going to be spent. An essential aspect of the evaluation process is to see what kind of budget the management team produces. Some of the questions we ask are: Where is the money going to lead the company? For instance, is it going to achieve certain clinical or research milestones? Does it move the company closer to getting FDA approval to start the clinical trial? Will the money get the company to a significant value inflection point which will support the company's next financing round or solicit partnering interests or an exit? Another essential aspect of the evaluation process is to examine the proposed team. Who are the people? What are their qualifications? Are there too few or too many people for the job? Does the company need a CEO? Before we invest in a company, we usually want the company to identify an acceptable CEO, or else we will agree to hire a CEO after the financing.

Personnel issues are also an important element of negotiation. We often insist that the venture partners and syndicate members have to have some say in the company to be able to guide the company as it grows. Additionally, a company may insist on negotiating a separate set of terms for certain senior executives such as the CEO or president and other key employees such as the founder of the company. I personally prefer that the company adopt a fairly consistent benefits package across the company so that everyone in the company is treated equally. In the long run, this will promote collegiality and augur well for the future success of the company.

CLS has a fairly standard deal structure and we rarely deviate from one investment to another. Some of the trickier situations and potential deal breakers are when a company insists on an unrealistic valuation, a larger than necessary option pool, or a stringent anti-dilution clause.

Using a Unique Business Model

We, though, do sometimes deviate from our standard deal structure in cases that involve a particularly promising disruptive technology. If we have a founder with a unique technology and we have confidence in the quality of the team, we will formulate a strategy to bring the company to the point where they can raise a Series A. For example, we recently invested in an early stage company in Singapore that developed a proprietary technology to create bioabsorbable polymeric stents. We funded the company and brought it to the U.S. to pull together a management team and infrastructure to conduct U.S. clinical trials. With this arrangement, we were able to take advantage of a low pre-money valuation on a unique technology while maximizing its potential through the exploitation of our firm's clinical development expertise in the U.S. This business model example is also of significant interest to global companies because it provides an avenue for them to access the U.S. capital market.

Misguided Entrepreneurs

Venture capitalists have many reputations and in some circles it is not a positive one. Some entrepreneurs may have an image of the VC as a "vulture capitalist" someone who is only there for the financial gain. Although VCs must endeavor to get good returns for their investors, VCs also have an interest in the technology and derive significant gratification from having played a part in benefiting humankind. This particular value is frequently underestimated and unrecognized by entrepreneurs. Furthermore, VCs have the common sense to understand that an ownership in a failed company is worthless. When we decide to make an investment in a company, we have made a determination that the entrepreneur's goals for the company is in line with our goals and that the deal terms are fair to ensure a productive relationship.

Frequently an inexperienced founder or an academician falls in love with their science and technology and has not taken into consideration the numerous risks of development. In these cases, one of the most common mistakes of entrepreneurs is to over-value their company. Entrepreneurs sometimes look around the industry and value the company significantly more than what the VCs know is the acceptable valuation. In some cases, the entrepreneur will forgo VC funding and strike a deal with an industry strategic investor. The terms may be more attractive and provide for an up-front payment and an attractive valuation. Sometimes, due to new corporate mandates and not necessarily due to any shortcomings of the company, the industry partner will back out. The company is now left in a difficult position where subsequent financings for the company will be a painful and unpleasant process. A good relationship and mutual understanding with the corporate partner is a key to success. One caveat is that there are a number of experienced entrepreneurs who have their own access to financial resources and can develop the company to a late stage position without the need for institutional funding.

The biotech industry is evolving with the venture community. The life science sector and biotech investing in general, although growing steadily, are still not well understood. Many investors still consider an investment in this area to be "high risk" due to the complexity, long incubation and development timeline and the highly specialized regulatory and reimbursement challenges. Nevertheless, with a worldwide aging population in the industrialized and developing regions of the world, there is an urgent and critical need for novel diagnostics and treatment for public health problems as well as intractable diseases. I believe the engine to propel this evolution will have to come from multiple directions, including universities, government agencies, private foundations, pharmaceutical companies, and venture capitalists. I believe early stage biotech investors are the liaison and accelerator of this dynamic research, development, and commercialization process.

Dr. Nelson Teng, Managing Partner of Charter Life Sciences, has had a very successful and active medical and scientific career. Since 1986, he has been chief of gynecologic oncology at Stanford University School of Medicine. Dr. Teng is a named inventor on fifteen patents and is an author of more than one hundred publications. Over the past twenty years, the scientific research at Dr. Teng's laboratory at Stanford University has focused on cancer biology, immunology, and gynecologic diseases. His laboratory has created numerous monoclonal antibodies of clinical interest, including the first human therapeutic monoclonal antibody to be taken into clinical trials. At Stanford, Dr. Teng has also been a principal investigator for more than thirty cancer clinical trials.

Among his many professional honors, he has been honorary senior registrar at Hammersmith Hospital (London), Royal Postgraduate College, research scientist at the Massachusetts Institute of Technology, gynecologic fellow at Stanford University School of Medicine, fellow at the American College of Surgeons, fellow at the American College of Obstetricians and Gynecologists, president of the Western Association Gynecological Oncology and chairman of the uterine and cervical cancer committee and member of both the ovarian cancer committee and the cervical cancer screening committee of the National Comprehensive Cancer Network. Dr. Teng has also served on various committees within the Gynecological Oncology Group, a national clinical trial consortium sponsored by the National Cancer Institute.

In addition to his many professional accomplishments, Dr. Teng has also been an active entrepreneur and investor over the last twenty years. He has been a founder of several companies including women's health care company Adeza Biomedical (ADZA), biotech company Univax Biologics (IPO, then acquired by NABI), biotech company SeraCare Life Sciences (SRLS), biotech company Palingen and women's health care company Vesta (acquired by Pfizer).

He has been a scientific advisor to Charter Venture Capital for more than fifteen years and a senior consultant to Mingly Capital. He has also been an advisor to Asian governments and other venture capital funds with respect to biotech matters.

Dr. Teng received his Ph.D. in biophysics from the University of California, Berkeley, his M.D. from the University of Miami, and is a graduate of a postgraduate program of the Stanford University Graduate Business School.

Acknowledgments: *Fred Schwarzer, Barr Dolan, Gino Di Sciullo, Jean Kanerva*

Paving the Venture Capital Road to Success in Biotech

Peter N. Allison

Managing Director

Biogenesis Investment Group LLC

Generating Success in Biotech

There is a great deal of cool science out there, but whether or not that translates into a business model for a drug has to do with looking at market share and real valuation. For instance, there are more than thirty cholesterol drugs on the market. If you were to invent another cholesterol drug and spend a lot of money getting it approved, you would get a very tiny share of market at the end of the day. On the other hand, there are drugs like Viagra and Cialis, which dominate their market. Going after markets where there are very few drugs is extremely lucrative. Much attention must be spent on analyzing the market. The market can be defined as the probable unit sales times the price. Sometimes a rare disease therapy will get orphan drug status (less than 100,000 cases, be cheaper to get approved, and sell for a very high price). Since it costs between $250 million and $1 billion to develop a drug you must do your homework on the market if you are going to get a return on your capital.

An awful lot of money is spent on overhead in the early stages of a company when there is not a lot of overhead required. Too many biotech companies start out with a scientist and technology and then they add businesspeople to the equation. They bring in a CEO and a CFO to raise capital and format the company, but mostly those people just stand around and attend biotech shows and meetings. We are often approached by small biotech firms. When you look at their cash requirements in their business plans, you can see that a lot of money goes to overhead for businesspeople. In the early stages of drug development, for at least a couple of years, you really don't need to have that group involved.

Venture capital companies will begin to provide more and more management services so that they can take on a group of new technologies. Instead of investing $5 million into a company where $2 million goes into actual drug development and $3 million goes into G&A and salaries, they will be able to invest in several different drugs, do the management and bookkeeping internally, and move the drugs up into their development process with the FDA before they start adding a lot of business overhead. By doing that, they will be spreading their risk and capital over several drugs, and they will be able to pre-emptively cull out the ones that aren't going to make it. Once a drug gets into the Phase II arena, then the company and drugs are more likely to attract an industry partner that will help pay the costs of further development, so it's a much better use of capital.

Making Successful Investments in Biotech

In order to determine successful investments, we investigate the market costs of drugs and also what it costs to get there. Some drugs can be moved forward quickly because a great deal of work has been done at the research institution or university. More and more VCs just aren't getting involved in the early stages. They are expecting that the work gets done through the grant process. We want to see a drug moved to Phase II very quickly with a minimum amount of investment. That is an important key to success. The second rule is to partner early and often. The quicker you can bring in an industry partner, the better. If you have a pulmonary drug in development and find a large company that does pulmonary work and quickly attract them as a partner, that is ideal. I would rather have a 10 percent interest in hundred oil wells than a 100 percent interest in ten oil wells. We like to spread our risks over as many technologies as we can. Early partnering can also provide mentoring, share in the development costs, and do the marketing work, as well as investing in the biotech startup.

We tend to be an earlier stage investor than most. We do a lot of work with the universities and the research institutes looking for promising technologies in medical devices. Most people think of this as being in the early stage of development, but there is in fact quite a bit of work being done. We chose this arena because it wasn't very crowded and we could spread our capital over several technologies because they were less expensive to license than later stage deals.

Spotting Opportunity in the Industry

We do a great deal of comparative analysis in order to pinpoint opportunities in the biotech industry. That means we look at opportunities and use modeling tools to get the technologies netted down to the ones that fit all the criteria: minimizing the costs to get the drug into Phase II, the likelihood of getting an industry partner, and the quality of the science and scientists. We tend to provide the management services, and we will invest in the technology and scientists, but we won't invest in overhead surrounding the business. The biggest thing we do is to enforce early stage cost control.

A Different Approach

We place a heavy emphasis on valuation. Additionally, we have built a very strong mathematical model of predicting risk and adjusted net present value of drugs once they hit the market. Therefore, we do a great deal of mathematical analysis and use predictive models to show what drugs are going to be worth. We don't take drugs all the way to market; we sell them in their development stages. Many companies get to second, third, or fourth round investments with the idea that they are not only going to invent and develop the drug, but they are also going to market the drug. I believe that is a big mistake. The biotech industry is good at inventing drugs and managing them through the FDA process, but typically, biotech companies have failed when they have attempted to get into big-time marketing of the drug. It is necessary to sell the drug to a much larger company when it is in late stage Phase II or about to enter Phase III. Our group insists that we are not financially involved all through the marketing.

Given the statistical probability of drugs being approved in the FDA formula, we are looking for at least a ten to one return in a relatively short time because there are so many failures. We try to factor fairly high internal rates of return, but the predictor models assume that there will be a certain failure rate in the drug portfolio.

Assessing Risk

We do a rigorous analysis of the science of a drug in order to decide whether the risk is worth taking. We don't invest in things that might work; we want to see some kind of human data and some kind of real potential. We also do very rigorous patent analysis to ensure that there aren't blocking patents out there and that the IP is properly protected. Then we fall back on cost control in the early stages. Hopefully, these valuation models tell us that these drugs are going to be worth a great deal of money once they are approved and that they won't be one of thirty other drugs in the same category.

Challenging Areas

Unlike software and telecommunications, which are typically good VC type investments, when you are dealing with a regulatory agency like the FDA and human life, the approval process is horrific. If you develop software that has a lot of glitches and doesn't work right, you will sell a little bit and then people will find out that it doesn't work and they won't buy it anymore. If you do that with a drug, you will kill people. The rigorous approval process is what distinguishes biotechnology from all other VC.

The Ideal Management Team

In pulling together a management team, we look for real dedication to moving the drug or the device through the approval process. We don't worry about good accounting or things like that because we can provide that service. We don't look for general business acumen. Instead, we focus on seeking out teams that are capable of working with the FDA and improving the science. They should be very focused on getting their product through approval because too many scientific groups tend toward being involved in research and are consequently unfocused. That work is fine, but it needs to be done at research institutes and funded by NIH and other grant organizations. Grants are for experiments, early stage science, and general science. Venture capital has to be moving toward an outcome.

Common Management Mistakes

Often management teams make the mistake of not being focused on getting their technology approved. They concentrate too much on doing more research and more product development as opposed to seeking product approval. Sometimes they tend to load up on business management when they don't need it, as the primary function of a company is to work with the FDA. You need FDA people, science people, and FDA compliance people. You do not need CFOs, finance people, or marketing people. Those functions aren't going to be relevant for five years. Unfortunately, some teams stay with a technology after it is evident that it won't succeed. You need to cut your losses and go find another project.

VC Mistakes

Sometimes companies continue to fund projects that have probably outlived their real ability. In other words, they start funding and stay with the project for too long. The ideal scenario is to be honest with oneself as to the time value of money and cut your losses. The second common mistake is that other companies fund a great deal of the infrastructure. When companies fund huge overhead and a hundred employees, it isn't relevant to the process. The single biggest mistake in the industry is allowing scientists to venture into the business world and create huge overheads. This chews up valuable research dollars that could be spent on the FDA approval process.

Developing a Good Track Record

The ideal scenario is to select good science, advance that science through the approval process, and sell it at a profit to another pharmaceutical company at the right time for a multiple of what you spent on development. Being able to get out at the right moment is probably the single biggest skill. It is essential not to wait for the company to turn into what is referred to as a FIPCO (fully integrated pharmaceutical company). Too many dollars are lost by not selling these technologies to larger companies. Big pharmaceutical companies own the retail markets. Biotech companies, and their investors, need to be focused on developing early stage and mid-stage drugs and selling them. It is crucial to have a successful early exit strategy. There is a role for federal grants in the early stage, and then there is a role for VC, followed by the large pharma investors. When a VC comes too early or stays too long, that is when losses rack up.

Exit Strategy

A good exit strategy is about more than just timing. It is necessary to develop a strategy and stick with it. You must determine that you are going to sell at a particular multiple and that you are not going to invest until there is a certain amount of information on the drug or a certain probability of success. You must have the discipline to get out at the right time and move on to the next deal. Trying to be a little bit of everything tends to get investment companies into trouble. The second piece of the puzzle is that the VC company must decide early on whether they are going to be the lead investor or if they are going to do group investing, whether someone else leads it or if you are going to be a late stage investor or follow-on investor. Those decisions are really important.

The Structure of the VC Deal

Typically, VC deals are structured where the company gives preferred stock to the venture capital firm. They will take anywhere from 40 to 60 percent of the value at the outset. If things don't work out right, they will be able to liquidate. Somebody once described the process as two pages of what we are going to do and fifty pages of what we will do if it doesn't work out. There is no real standard model because there are many different variables. One major factor is who controls the underlying IP on a technology or device. Much of it has to do with making certain that your IP is protected and that any patents that come up while you are developing or any patent protection that needs to be done is paid for. There should always be provisions for more money to come in.

Make sure you have a clear understanding of what happens if milestones aren't met because all VC works on milestones. They will fund for certain things to be accomplished, and if those things aren't accomplished, then it is important for the founders to understand what happens next. If you do hit your goals, you must determine whether there will be additional funding or not. With most deals, a lot of time is spent on those topics. Founders tend to take money wherever they can get it and tend to think that once they get the money, they can do what they want. That is simply untrue. You have a very serious engagement with whoever is doing your funding, and it is very important that you share management and business philosophy. You must be aligned for the same goals and outcomes. It is crucial that companies and founders like each other because they will have to work very closely for a long period of time.

Negotiations with the Entrepreneur

The direction of a deal ultimately comes down to control. It is about who controls what. From the founder's standpoint the fears and bad reputation that VCs have are widespread, often because founders are not clear about the control of the enterprise. It always comes down to whether the money or the scientists are going to control things. Typically, the solution is to ask the scientists to write out their plan as to how it's going to work. As long as they proceed according to plan and do a good job and hit important milestones, everything will be fine. Problems spring up when the milestones aren't getting hit and something comes up that requires more money. The VC might declare that it is time to shut down the company. Founders and scientists are willing to do anything to get the money, and once they get the money, they think they can do whatever they want. That disrupts deals and creates disharmony, and once that starts happening, it's just a matter of time before that enterprise fails.

Important Deal Components

The three most important parts of a deal are the ownership of the IP, the commitment to budgets, and the exit strategy. Once it clears Phase II and goes to Phase III, that is the point at which I want be sold. On the contrary, founders want to build a big company and hire salespeople. At that point, they are going to want to buy the VCs out, and the VCs will want to have a big return. These issues are best dealt with in the funding and shareholders agreements at the outset.

IP Negotiation

Negotiations take place in order to figure out how to dispose of the IP if things don't go right. This is crucial because the IP is probably the only asset the company has. If there are problems, at what point do the founders get to take the IP and go off and try again? Does the VC firm have the right to take the IP and sell it in order to recoup their money? This needs to be decided from the beginning. Again this should be covered in the early documents with reasonable give and take.

Budgets

There needs to be an understanding that overhead can increase as the value of the drug increases. As you move through the drug development process and the drug develops a much higher value and the risks of approval go down, you want to tie your overhead to that increased value. There should be a strict adherence in the early days to very tight budgeting and very low overhead, except as it deals with approval of the drug.

Room to Negotiate

Because we don't spend a lot of money on business overhead, we like science founders to look at us as providing the business acumen in the deal. The problem is when you get scientists who think they are also businesspeople. You should be able to trust that the scientist is committed to getting his science approved and not doing general research. In turn, the scientist should trust the investment company to handle the business affairs of the company. This should yield a successful partnership. When people cross over into areas that they are not familiar with, there will be trouble. Early on, we spend the most time on discussion and education rather than negotiating. The VC firm has to understand the strengths and weaknesses of the technology as it relates to the FDA approval process, and the scientists have to understand the business side of the equation. If we are going to invest X amount of dollars, then it has to get the company from point A to point B.

Different Deal Structures

In an early stage deal, it is important to get validation very quickly. You want to spend the least amount of money to get the most validation possible. When you look at an early stage deal incorporating human or animal toxicology, you want that done at a very reputable lab and very quickly for the least expense possible. This is because when you are in an early stage drug development deal, the likelihood of it failing is very high. Therefore, you don't want to spend a lot of money on it. Often, we will engage in a two-step process. The first step will be to take X amount of dollars and validate the science. If it is okay, the next stage is the real deal. You may just take an option on the drug or technology and set up the ultimate deal then, but instead of going into a deal, you go into an interim relationship. If you are going to do the entire deal yourself, you have more latitude in how you structure it. However, if you are expecting otner firms to join in on your investment, then they have to recognize the structure.

Deviating from the Norm

The rules go out the window when you have a superstar in the science world. Dr. Leroy Hood has been hugely successful in the biotech business. He dictates the terms and the probability of being successful at something he wants to do. Investors line up to get into one of his deals. The probability of success is very high compared to a scientist in a university without a reputation. It is pretty rare to find such a superstar. Dr. George Rathman, a founder of Amgen, is another who went on to found several successful companies as did Dr. Steven Gillis, a founder of Imunex who then built Corrixa, and now is a partner in a very large VC firm.

The biotech industry holds the hope of curing all currently known diseases in the next fifty years. It will change medicine as we know it today. But that will require an enormous amount of money and the industry has not provided a very good return compared with, say, telecom, software or the Internet. Money is a commodity like oil and wheat. It goes where it can get the highest price/return. We need to adopt better government grant and education funding for early stage research in order for the U.S. to remain a leader. Today we have the finest research universities and institutes in the world. We can only continue that stature with adequate and increased funding for the future.

Peter Allison has more than thirty years of experience in real estate development, investment, and venture capital. He is the president and CEO of IDM Technologies Inc. and the manager of BioGenesis Investment Group LLC. His passion is biotechnology with special interest in how critical diseases are treated, sequencing the human genome, and adult stem cell research. Mr. Allison has a B.A. in economics from the University of Washington and a J.D. with concentration in corporate securities and tax law from the University of Puget Sound. He is very active in numerous professional associations and volunteer organizations and sits on two university boards.

APPENDICES

Appendix A

COMPANY X
SERIES B PREFERRED STOCK INVESTMENT
TERM SHEET

Closing Date: June 5, 2006

Issuer:	Company X (the "Company")
Type of Security:	Series B Convertible Preferred Stock (the "Preferred")
Investment Amount:	$30 million

Purchase Price per Share:	$1.25
Use of Funds:	General working capital in support of the operating milestones mutually agreed by the company's senior management, the Board and the Investors.
Dividends:	The Preferred will earn non-cumulative dividends at a rate of 8% per annum in preference to other series of preferred stock and the Common Stock when, as and if declared by the Board of Directors.
Conversion Rights:	Each holder shall have the right, at any time, to convert the Preferred shares into the Company's Common Stock at a Conversion Price equal to the Purchase Price per Share.
Automatic Conversion:	The Series A and Series B Preferred will automatically convert to shares of Common Stock upon (i) the approval of at least 66 2/3% of the Series A and Series B Preferred holders voting together, or (ii) upon the closing of an Initial Public Offering (IPO) of Common Stock with a per share price of no less than 3 times the Conversion Price and aggregate gross proceeds of at least $30 million. The Conversion Price shall equal the Purchase Price per Share.
Redemption Option:	On or after seven years from the date of issuance, the Series B Preferred will be redeemable in whole and not in part, at the option of the holders of a majority of the

then outstanding shares of such series, at a value equal to the Purchase Price per Share (as adjusted) plus any accrued and unpaid dividends thereon to the redemption date payable. If redemption is requested, payment may be made in three equal annual installments with the first payment due within 180 days of the request. The Investors' Redemption Option shall be contingent upon the Company having legally available funds to make such payment.

Liquidity Event, Liquidation and Liquidation Preference:

In the event of a Liquidity Event (defined below), the holders of the Series B Preferred will be entitled to receive, in preference to the holders of the common stock and other outstanding series of preferred stock, an amount equal to one times the then applicable Purchase Price per Share plus any accrued and unpaid dividends (the "Series B Liquidation Preference.

After payment of the Series B Liquidation Preference, the holders of the Series A Preferred will be entitled to receive, in preference to holders of the common stock, their applicable liquidation preference plus any accrued and unpaid dividends (the "Series A Liquidation Preference").

After payment of the Series B Liquidation Preference and the Series A Liquidation Preference, the remaining assets or property distributable upon such liquidation shall be divided pro rata among the holders of the Common Stock and the holders of the Series B Preferred on an as-converted basis until all amounts received by the holders of the Series B Stock (including the Series B Liquidation Preference) equal 3 times the original Purchase Price, after which any remaining proceeds of the Company shall be distributed among the holders of the Common Stock.

A Liquidity Event shall be defined as (i) a sale of all or substantially all of the Company's assets, (ii) a consolidation or merger of the Company in a transaction in which the shareholders do not continue to hold at least 50% interest in the successor entity (or in which the Company is not the controlling entity), (iii) a change in ownership in any 12 month period of more than 50% of the outstanding Common Stock of the Company, or (iv) a liquidation, dissolution or winding up of the Company or its assets; provided, that an equity financing transaction shall not be considered a Liquidity Event.

Anti-Dilution Protection:

The conversion price of the Series B Preferred will be subject to broad-based weighted average Anti-Dilution protection.
The conversion price will also be subject to proportional adjustment for stock splits, stock dividends, re-capitalizations and the like.

Priority:	The Series B Preferred will be senior to all other classes of stock in terms of rights to distributions relating to redemption, dissolution, liquidation or winding-up of the Company and the declaration and payment of dividends.
Pro Rata Rights of First Refusal (Pre-emptive Rights):	The holders of at least 300,000 shares of Series B Preferred shall have the right of first refusal to purchase up to their pro rata share (based on its percentage of the Company's outstanding common shares, calculated on an as-if-converted basis) of any equity securities offered by the Company, subject to carve-outs typical in such transactions.
Shareholders Agreement:	The Shareholders Agreement shall contain customary covenants, representations, warranties and indemnities (or, as appropriate, these may be incorporated into the other transaction documents). Without limiting the generality of the foregoing, the transaction documents will also include certain protections for the Investors such as: tag-along rights; drag-along rights; approval of the Preferred for acquisitions; and other appropriate protections (see Protective Provisions).
Voting & Protective Provisions	The Series B and Series A shall vote together as one class on all matters except as specifically stated below: The Company will not, without consent of the holders of at least 66 2/3% of the outstanding Series A

and Series B Preferred, voting together as a single class (i) merge or consolidate into or with any other corporation or sell all or substantially all of the Company's assets; (ii) liquidate, dissolve, or wind up the Company, (iii) amend the Company's Bylaws or Articles of Incorporation; (iv) reclassify any Common Stock into shares having preference equal to or over the Preferred; (v) incur new indebtedness in excess of $500,000 (excluding: a) asset-backed debt in connection with Board-approved transactions; b) seller notes as consideration for Board-approved acquisitions, provided such notes are unsecured obligations of the Company, or c) capital lease obligations incurred in the ordinary course of business); (vi) pay or declare any dividend or redeem, repurchase or acquire any shares of its capital stock (except for shares repurchased in accordance with restricted stock purchase agreements with employees, consultants, or directors which have been approved by the Board); or (vii) alter the size or composition of the Board of Directors (unless unanimously approved by the Board).

At least 66 2/3% of the Series B Preferred shall be required to authorize or issue any new or existing class or series of capital stock having a preference over, or being on parity with the Series B Preferred;

The Series B Preferred and Series A Preferred shall vote separately on the following matters, with only the relevant Series voting as it relates to their particular class of stock. As applicable, 51% of the Series A and 66 2/3% of the Series B shall be required to (i) materially and adversely alter or change the rights, preferences, or privileges of that respective class of Preferred; (ii) increase the authorized number of shares of the respective Series of Preferred Stock; (iii) declare or pay any dividends or other distributions to holders of junior stock; or (iv) redeem or purchase for cash any capital stock of the Company that is junior to or pari pasu with the respective Series of Preferred Stock (except for shares repurchased in accordance with restricted stock purchase agreements with employees, consultants or directors, with Board approval).

Registration Rights:

The Preferred holders will be granted customary registration rights, including at least 2 demand rights and unlimited S-3 and "piggy-back" registration rights. The Company will not grant registration rights which are superior to or which would interfere with the Preferred's rights without the consent of a majority of the holders. Costs of effecting of such registration(s) shall be borne by the Company.

Board of Directors:	The Board of Directors shall consist of 7 members divided among the shareholding contingencies in the following manner: 1) Series B Preferred: 3 designees: • Investor A shall have the right to appoint one designee, who shall initially be John Doe. • Investor B shall have the right to appoint one designee, who shall initially be Jane Smith. • Investor C shall have the right to appoint one designee, who shall initially be Fred Wright 2) Series A Preferred & Common: 1 designee selected by the Series A and Common Shareholders, voting together. The Series A/Common designee shall initially be John Doe. 3) Chief Executive Officer: 1 designee, who shall be the Company's CEO. 4) Independent Board Members: The Board will initially nominate and elect

2 independent members to the Board; such member will be identified by management and will be reasonably acceptable to the Series B investors.

The Board will establish and maintain Compensation and Audit Committees consisting of non-Management designees. The Investor A designee shall have the opportunity to serve on the Compensation, Audit or any other committee of the Board of Directors. The Board will meet at least monthly unless otherwise mutually agreed by its members.

Information Rights

The Company shall deliver to the Investors holding at least 300,000 shares of Series B Preferred the following information:

1) Within 30 days after the end of each calendar month, detailed monthly financial statements including consolidated income statement, balance sheet and statement of cash flows

2) Within 45 days after each fiscal quarter, quarterly consolidated financial statements for the Company

3) Within 90 days after each fiscal year, annual consolidated financial statement for the Company

4) Within 180 days after each fiscal year, audited financials for the Company prepared by an audit firm approved by the Board.

5) Within 30 days following the start of each fiscal year, a detailed monthly financial forecast.

6) On a timely basis, all documentation reasonably required to ensure Investors' ongoing compliance with rules and regulations governing an SBIC fund.

All financial information provided by the Company will be reported in accordance with GAAP, except for minor variances in unaudited monthly financial statements.

Management Option Pool and Warrant Pool:

Company X's post-money capitalization will include a total pool of common stock for employee options, strategic warrants, stock grants and the like. For a period of 12 months post-Closing, any new Options created in addition to the New Option Pool shall not be dilutive to the Series B holders.

Company Right of First Refusal for Sales of Management Shareholder Stock

In the event a Management shareholder desires to sell or transfer his/her Common Stock, the Company shall have a right of first refusal in connection with such sale or transfer. Should the Company decline to purchase such shares, the Preferred B and

	Preferred A holders shall then have the right of first refusal to purchase such shares on a pro rata basis.
Deal Expenses:	If the Closing occurs, the Company shall reimburse the Series B Investors for all reasonable due diligence and legal expenses related to this investment upon the Closing, not to exceed $40,000 of actual total expenses. The Company's counsel shall take the lead in the drafting of definitive transaction documents. For purposes of simplicity, such reimbursement amount shall be withheld from the final Closing amount and the Investors shall provide the Company with receipts and documentation necessary for tax purposes pursuant to these expenses.

Courtesy of Heidi L. Huntsman, UV Partners

Appendix B
Capital Structure for Company XYZ

	Amount Invested	Shares	% Ownership	Return Scenario $80 million	Return Scenario $200 million	Return Scenario $400 million
Common Share-holders*		70,000,000	70%	$35	$119	$280
Investor 1	$10,000,000	10,000,000	10%	$15	$27	$40
Investor 2	$20,000,000	20,000,000	20%	$30	$54	$80
Total		**100,000,000**	**100%**	**$80**	**$200**	**$400**

***Both investors have a 1x preference with a 3x cap on participation**

Courtesy of Heidi L. Huntsman, UV Partners

ASPATORE BOOKS

C-LEVEL BUSINESS INTELLIGENCE • C-LEVEL BUSINESS INTELLIGENCE •